ELECTROANALYTICAL CHEMISTRY

VOLUME 15

ELECTROANALYTICAL CHEMISTRY

A SERIES OF ADVANCES

Edited by
ALLEN J. BARD

DEPARTMENT OF CHEMISTRY
UNIVERSITY OF TEXAS
AUSTIN, TEXAS

VOLUME 15

CRC Press
Taylor & Francis Group
Boca Raton London New York

CRC Press is an imprint of the
Taylor & Francis Group, an **informa** business

First published 1989 by Marcel Dekker, Inc.

Published 2019 by CRC Press
Taylor & Francis Group
6000 Broken Sound Parkway NW, Suite 300
Boca Raton, FL 33487-2742

© 1989 by Taylor & Francis Group, LLC
CRC Press is an imprint of Taylor & Francis Group, an Informa business

First issued in paperback 2019

No claim to original U.S. Government works

ISBN-13: 978-0-367-45121-9 (pbk)
ISBN-13: 978-0-8247-7646-6 (hbk)

Visit the Taylor & Francis Web site at
http://www.taylorandfrancis.com

and the CRC Press Web site at
http://www.crcpress.com

The Library of Congress Catalogued the First
Issue of This Title as Follows:

Electroanalytical chemistry: a series of advances, v. 1-

  New York, M. Dekker, 1966-
      v. 23 cm.
  Editor: 1966- A. J. Bard

   1.Electrochemical analysis-Addresses, essays, lectures.
  1. Bard, Allen J., ed.

QD115E499   545.3   66-11287

Library of Congress
ISBN 0-8247-7646-1 (v. 15)

Introduction to the Series

This series is designed to provide authoritative reviews in the field of modern electroanalytical chemistry defined in its broadest sense. Coverage will be comprehensive and critical. Enough space will be devoted to each chapter of each volume so that derivations of fundamental equations, detailed descriptions of apparatus and techniques, and complete discussions of important articles can be provided, so that the chapters may be useful without repeated reference to the periodical literature. Chapters will vary in length and subject area. Some will be reviews of recent developments and applications of well-established techniques, whereas others will contain discussion of the background and problems in areas still being investigated extensively and in which many statements may still be tentative. Finally, chapters on techniques generally outside the scope of electroanalytical chemistry, but which can be applied fruitfully to electrochemical problems, will be included.

Electroanalytical chemists and others are concerned not only with the application of new and classical techniques to analytical problems but also with the fundamental theoretical principles upon which these techniques are based. Electroanalytical techniques are proving useful in such diverse fields as electro-organic synthesis, fuel cell studies, and radical ion formation, as well as such problems as the kinetics and mechanisms of electrode reactions, and the effects of electrode surface phenomena, adsorption, and the electrical double layer on electrode reactions.

It is hoped that the series will prove useful to the specialist and nonspecialist alike—that it will provide a background and a starting point for graduate students undertaking research in the areas mentioned, and that it will also prove valuable to practicing analytical chemists interested in learning about and applying electroanalytical techniques. Furthermore, electrochemists and industrial chemists with problems in electrosynthesis, electroplating, corrosion, and fuel cells, as well as other chemists wishing to apply electrochemical techniques to chemical problems, may find useful material in these volumes.

A.J.B.

Contributors to Volume 15

H. H. J. GIRAULT* Department of Chemistry, University of Southampton, Southampton, England

SHIMSHON GOTTESFELD Electronics Research Group, Los Alamos National Laboratory, Los Alamos, New Mexico

D. J. SCHIFFRIN Wolfson Centre for Electrochemical Science, University of Southampton, Southampton, England

R. MARK WIGHTMAN Department of Chemistry, Indiana University, Bloomington, Indiana

DAVID O. WIPF Department of Chemistry, Indiana University, Bloomington, Indiana

*Current affiliation: Department of Chemistry, University of Edinburgh, Edinburgh, Scotland

Contents of Volume 15

Contents of Other Volumes

VOLUME 1

VOLUME 2

ELECTROANALYTICAL CHEMISTRY

VOLUME 15

ELECTROCHEMISTRY OF
LIQUID-LIQUID INTERFACES

H. H. J. Girault*

*Department of Chemistry
University of Southampton
Southampton, England*

D. J. Schiffrin

*Wolfson Centre for Electrochemical Science
University of Southampton
Southampton, England*

---

*Current affiliation: Department of Chemistry, University of Edinburgh, Edinburgh, Scotland.

1

I. INTRODUCTION

The electrified interface is an important aspect of all heterogeneous chemical systems. It is a vital feature of many biochemical systems as well as of other less complex systems, including, for example, colloids, gels, artificial membranes, and metal-electrolyte interfaces. The latter has been the subject of intense study and much of the language and hence our understanding of the electrified interface stems from experimental studies on metal electrodes, especially at the mercury/aqueous electrolyte interface.

The success of these studies derives from several factors but especially from the smoothness, the unreactivity, and therefore the reproducibility of this interface. Attempts to carry over some of these certainties into other systems, such as colloids, gels, and membranes, immediately meet with the problem of the unknown physical structure of their surfaces and even of their unknown chemical composition. Such drawbacks do not, however, attach to another type of interface—the interface between two immiscible electrolyte solutions (ITIES) neither of which is a metal.

II. HISTORICAL OVERVIEW OF THE PHYSICAL CHEMISTRY OF ITIES

A. Concentration Cells and Equilibrium Potentials

The first direct electrochemical study of ITIES was due to Nernst and Riesenfeld [1[in 1902. Using colored inorganic electrolytes [KI_3, K_2CrO_4, $Fe(SCN)_3$, etc.] at partition equilibrium between water and phenol, they observed the transfer of ions during the passage of current through the system water/phenol/water.

The aim of the work of the school of electrochemistry in Göttingen at the beginning of the twentieth century was primarily to find methods to measure transport numbers in nonaqueous solvents. In their first approach to calculating the transport number of colored salts in phenol, Nernst and Riesenfeld quantified the observed phenomena, taking into account mass transfer by both diffusion and migration. Assuming that the transport number of the cation was constant for each phase, they calculated a transport equation at constant current from the condition of equality of the flux of cation J_+ at the interface (X = 0) given by [2]

$$J_+ = -D_x^w \left(\frac{\partial c_+^w}{\partial X} \right)_{X=0} + \frac{z t_+^w}{F} j = -D_+^p \left(\frac{\partial c_+^p}{\partial X} \right)_{X=0}$$

$$+ \frac{z t_+^p}{F} j \tag{1}$$

where t_+, D_+, and c_+ are the transport number, the diffusion coefficient, and the concentration of the cation; j is the current density; w and p refer to the aqueous and the phenol phases, respectively.

To calculate t_+^p as a function of t_+^w, Nernst and Riesenfeld solved, independent of the previous work of Sand [3], the Fick diffusion equation to express the interfacial gradient of concentration of the cation as a function of the bulk concentration. They concluded that the electrolysis of the cell

| aqueous electrolyte | | phenol | | aqueous electrolyte |

where the ionic concentrations are the same in the two aqueous compartments, could provide a means for measuring the transport number of an
ion in the organic phase from its value in aqueous solution.

The second approach used by Riesenfeld to measure transport numbers
in organic solvents was based on the measurement of the electromotive
force (emf) of concentration cells with organic liquid membranes. In
1902, this author [4,5] showed that although it is not possible to measure
directly the difference of potential across the water-phenol interface, the
emf of concentration cells such as

| electrode | aqueous electrolyte concentration C_1 | phenol | aqueous electrolyte concentration C_2 | electrode |

could be expressed as a function of the aqueous electrolyte concentrations
according to a Nernst equation,

$$E = \frac{2RT}{F} t_-^{\,p} \ln \frac{c_1}{c_2} \qquad (2)$$

This expression, used by Riesenfeld [5,6] to measure the transport
numbers of the electrolytes KBr, KCl, and KI in phenol, corresponds,
of course, to a concentration cell with transport, a fact that was not
recognized at the time. This second method to measure transport numbers
was absolute, as it is independent of the value of the transport number
in the aqueous phase, but has nevertheless been used very little.

After these preliminary electrochemical investigations, interest in the
ITIES spread to physiologists after Cremer [7], in 1906, pointed out the
analogy between the water/oil/water concentration cells and biological
semipermeable membranes studied by Ostwald [8]. The oil-water interface then became a model to study bioelectrical potentials and currents,
which had fascinated the scientific community since the first experiments
of Galvani in 1786.

During the first half of the twentieth century, several investigators
using different polar solvents, such as salicylaldehyde [9], o-toluidine

[9], or amyl alcohol [10], were carried out by other physiologists, including Beutner and Baur, in the hope that the understanding of these apparently simple systems would help to elucidate the more complex behavior of biological cell membranes.

The origin of the potential difference measured across a water/oil/water concentration cell when the concentration of various organic salts (e.g., tetramethylammonium salts or picrates) was varied in one of the aqueous compartments has been the subject of considerable controversy for many years since the potential differences were seen to be slowly changing during the course of the experiment.

At first, Cremer [7] claimed that the difference in the mobility between the cations and the anions was the origin of the emf of the cell. In modern terminology, this is equivalent to saying that the potentials measured were diffusion potentials.

Beutner [11—15], who initiated a systematic study of this type of concentration cells, believed that the emf observed arose from the free charge located at the oil-water interface due to an unequal distribution at the ions across the interface.

On the other hand, Baur [16—20] and later Ehrensvärd [21—25] thought that selective ionic adsorption at the oil-water interface was the cause of the phenomena observed and claimed that the potential differences measured were adsorption potentials or surface potential differences.

After a vigorous polemic reviewed in 1938 by Wilbrandt [26], Baur accepted Beutner's explanation of the origin of the emf resulting from concentration cells with liquid membranes. In 1940, Dean et al. showed theoretically [27—29] that Beutner's approach was correct and that the time for decay of the surface potential difference $\Delta\chi$ after the addition of a salt in one of the aqueous compartments of the cell determines whether the potential difference measured, $\Delta\phi$ (Galvani or inner potential), is a Volta potential difference, $\Delta\psi$ (outer potential), as Beutner suggested, or an adsorption potential difference $\Delta\chi$, as Baur indicated.

In 1953, Karpfen and Randles [30] subjected Beutner's approach to
a rigorous thermodynamic analysis and introduced the concept of distribution potential differences. This treatment, which was based on the
equality of the electrochemical potentials of ions in both phases, showed
that the potential differences due to the distribution of different salts
between water and diisopropyl ketone were independent of the salt concentration and were given by

$$\Delta\phi_{C^+A^-} = \frac{RT}{2F} \ln \frac{B_{C^+}}{B_{A^-}} \tag{3}$$

where C^+ and A^- represent the cation and the anion of the salt and B_i
is the ionic distribution coefficient (see Sec. III.C.1).

At about the same time, Bonhoeffer et al. pointed out the analogy
between water/oil/water concentration cells, where only one ion can
permeate the oil phase, and the water/ion exchange membrane/water cells
[31]. By applying to the oil phase the theory of Teorell [32] and Meyers
and Sievers [33] for fixed charged membranes, these authors were able
to study [34,35] the concentration gradient of the different ions within
the organic phase (quinoline) and to estimate the various components of
the membrane potential, which they divided into phase boundary and
diffusion potentials.

At about the same period (i.e., in 1955), Davies and Rideal [36]
were able, using a vibrating plate electrode [37], to measure separately
surface and diffusion potentials. They showed that adsorption potentials
$\Delta\chi$ can occur at the oil-water interface only if the oil is completely nonpolar (as in paraffin oils) and therefore concluded that Beutner's concepts were correct since the systems studied by this author were composed
of polar solvents, and interfacial double layers could be formed. Davies
and Rideal justified this difference due to the polarity and nonpolarity of
solvents by explaining the fact that $\Delta\chi$ was stable indefinitely with a
nonpolar oil due to the very low solubility of ionic species, which means

that no compensating double layer can be established in the thickness of oil available. The work of Davies and Rideal clarified the meaning of the different electrical potentials which characterize the electrified liquid-liquid interface and their physical basis.

These ideas were summarized in 1960 by Davies [38], who published a detailed and comprehensive review on the different types of potential differences occurring at the liquid-liquid interface (i.e, distribution, diffusion, and surface potentials). Later Dupeyrat [39,40] in 1964 and Gavach [41,42] in 1967 reviewed the different types of concentration cells, with liquid membranes showing especially the different analytical applications.

B.   Double Layer, Electrocapillarity, and Electroadsorption

Besides the different measurements of Galvani and surface potentials described in Sec. II.A, liquid-liquid interfaces were also investigated by colloid and surface chemists striving to obtain more information about the structure and the potential distribution at interfaces. In 1939, Verwey and Niessen [43] examined the distribution of charges and the potential function in the neighborhood of the interface. They proposed a model for the interface based on a back-to-back diffuse double layer with charges distributed in the two adjoining phases.

After the proposal of this model, the first experimental investigation of the interfacial structure was reported in 1954 by Kahlwert and Strehlow [44]. They showed that the potential drop across the water-quinoline interface was spread mainly over the two ionic double layers, but unfortunately the measurements were not accurate enough to estimate the potential drop due to a possible inner layer of oriented dipoles. In a later work, Strehlow [45] studied the water-ligroin interface (i.e., the interface between water and a nonpolar oil).

Taking as the aqueous electrolyte a mixture of anionic (sodium acetylsulfate) and cationic (cetylpyridinium chloride) surfactants, the surface potential could be changed by changing the composition of the mixture.

In this way, Strehlow constructed an electrocapillary curve point by point by measuring the surface potential with a vibrating electrode and the interfacial tension at constant area per molecule (1 nm^2/ion) with a ring tensiometer. The pioneering work of Strehlow was the first example of an electrocapillary curve for a liquid-liquid interface, in which the electrical state of the interface was established by ionic adsorption and distribution. A further analysis of Strehlow's work by Ohlenbusch [46] using the Verwey and Niessen [43] model showed that most of the potential drop occurred in the oil phase.

In contrast with the work of Strehlow, where the potential distribution is established by partition equilibrium, Guastalla [47] in 1956 studied the effect of externally applied potentials in the interface properties. This author [48] measured the steady-state interfacial tension between a solution of hexadecyltrimethylammonium bromide (0.5 mM) in nitrobenzene and an aqueous solution of potassium bromide (1 mM) as a function of applied potential. A decrease in the interfacial tension was observed when the potential of the oil was made positive with respect to the water; when anionic surfactant was employed the opposite effects were observed. The author concluded that these phenomena could not arise from classical electrocapillary effects, as the experiment gave interfacial tension changes only with surface-active ions. Guastalla called the observed phenomenon electroadsorption, as he regarded the interfacial field as the only cause of adsorption or desorption of ionic surfactants.

Similar steady-state mesurements were carried out in Japan by Watanabe and co-workers [49–56], who investigated a wide range of surface-active agents at the water-methyl isobutyl ketone, water-butanol, and water-pentanol interfaces. In particular, they studied the electroadsorption phenomena in the case of amphoteric surfactants such as lecithin and were able to detect the isoelectric pH as the pH for which no decrease in interfacial tension was observed with applied voltage.

The study of electroadsorption was also undertaken in 1963 by Blank and Feig [57], who attempted to make a distinction between electro-capillarity and electroadsorption. It was proposed that electroadsorption as described by Guastalla was a mass transfer effect. From this distinction, Blank [58,59] studied electroadsorption from a kinetic point of view by measuring the variation of interfacial tension with time when a constant current was applied across the liquid-liquid interface. It was shown that the observed decrease in interfacial tension, which was thought to denote the accumulation of ions at the interface, depended on the transport number of the ions, the ionic concentrations in both phases, and on the current density. It was concluded thus that electroadsorption was a mass transfer phenomena dominated by migration.

The electroadsorption effect was subsequently studied by Dupeyrat [60—63], Gavach [64,65], and Joos [66,67], all of whom used long-chain ionic surfactants and were able to confirm that the reasoning put forward by Blank was correct.

It is important to clarify in modern terms the concepts and ideas of electroadsorption discussed by Guastalla, Blank, and others in the 1960s. Using the now available data on Gibbs energy of partition of ions (see Table 1), it can be seen that for the original system studied by Guastalla [i.e., hexadecyltrimethylammonium bromide ($HTMA^+Br^-$) in nitrobenzene and KBr in water], when the oil phase is made positive with respect to water, the current is likely to be carried by Br^- across the interface and the concentration of the counterion in each phase will follow the concentration profile of Br^- to preserve electroneutrality. Figure 1 gives a schematic representation of a steady-state concentration profile for ionic diffusion and migration at constant current. From this figure it can be seen that the interfacial concentration of $HTMA^+$ is increased when the oil phase is made positive, thus causing a decrease of interfacial tension. On the other hand, when the oil phase is made negative, the current is likely to be carried mainly by the fast transfer of Br^- from oil to water.

TABLE 1

Gibbs Energies of Partition and Transfer[a] (kJ mole^{-1})

	Water/1,2-dichloroethane				Water-nitrobenzene			
	ΔG_p [1]	ΔG_p [2]	ΔG_t [3]	ΔG_c [4]	ΔG_p [5]	ΔG_t [6]	ΔG_p	ΔG_c [4]
				Cations				
Li$^+$					38.2			
Na$^+$			24.7	31.0	34.2			25.1
K$^+$			25.6	28.0	23.4	21.0		22.2
Rb$^+$			24.7	25.5	19.4	19.2		20.1
Cs$^+$		24.9 [k]	23.8	23.0	15.4	17.8	15.6 [j]	17.6
Mg^{2+}							69.6 [a]	
Ca^{2+}							67.3 [a]	
Sr^{2+}							66.0 [a]	
Ba^{2+}							61.7 [a]	
TMeA$^+$	17.6		15.4	12.3	3.4	4.0		13.0
TEtA$^+$	4.2		4.7	5.0	-5.7	-4.8		2.5
TPrA$^+$	-8.8		-8.8	-8.8		-16.4	-15.6 [b]	-10.0
TBuA$^+$	-21.8	-19.2	-16.4 [k]	-22.6	-24.0	-23.5	-24.7 [c]	-22.6
TPeA$^+$	-34.7	-33.1		-35.6				-35.1

Ion								
$THxA^+$	−47.7			−46.0			−45.6 [d]	−45.6
$THpA^+$		−45.2						
$T\Phi As^+$	−35.1	−32.5 [k]	−32.6		−35.9	−36.1	−31.4 [k]	
$T\Phi Ph^+$			−32.6			−37.3		
H^+					32.5			
NH_4^+					26.8			
$C_{10}TMA^+$	−13.4						−34.0 [e]	
$C_{12}TMA^+$	−20.2						−40.9 [e]	
$C_{14}TMA^+$	−27.1						<−44.7 [e]	
$C_{16}TMA^+$	−33.9						11.3 [f]	
Choline							4.8 [f]	
Acetylcholine								
$TBuPh^+$		−22.6						
CV^+							−39.5 [g]	
Anions								
F^-	56.1							
Cl^-	46.4	45.2	53.6	54.8	29.7	43.9	42.7 [h]	49.8
Br^-	38.5	35.6	39.3	39.7		36.0	34.3 [h]	34.7
I^-	26.4	23.8	25.1	22.6	18.8	21.9	20.5 [h]	18.0
SCN^-	25.5						10.0 [h]	
NO_3^-	33.9			35.6			23.8 [h]	30.5
ClO_3^-	33.1							

TABLE 1 Continued

| | Water/1,2-dichloroethane | | | | Water-nitrobenzene | | | |
	ΔG_p [1]	ΔG_p [2]	ΔG_t [3]	ΔG_c [4]	ΔG_p [5]	ΔG_t [6]	ΔG	ΔG_c [4]
Anions								
ClO_4^-	17.2		16.9	14.6	8.0	9.8	9.3 [g]	10.5
I_3^-					23.4			
I_5^-					38.8			
Octoate							−8.5 [g]	
Dodecylsulfate							4.1 [g]	
Dipicrylaminate					−39.3			
$T\Phi B^-$	−35.1		−32.6		−35.9	−36.1	−50.2 [i]	
Dicarbollyl cobalate				7.1				
Picrate					−4.6	−3.4		+3.3

[a]Data recalculated to present these values in kJ mole^{-1}, in accord with the $T\Phi As^+/T\Phi B^-$ Grunwald assumption from the following references: [1], Czapkiewicz [91]; [2], Antoine [92]; [3], Abraham [93]; [4], Abraham [84]; [5], Rais [94,95]; [6], De Namor [96]; [a], Marecek [97]; [b], Gavach [98]; [c], Gros [99]; [d], Abraham [84]; [e], Gavach [100]; [f], Vanysek [101]; [g], Vanysek [102]; [h], Gerin [103]; [i]. Koryta [104]; [j], Hundhammer et al. [184]; [k], Hundhammer and Solomon [188].

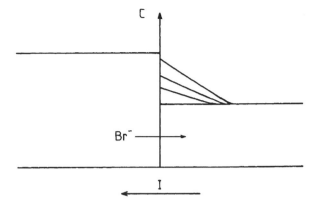

Aqueous KBr   HTMABr in Nitrobenzene

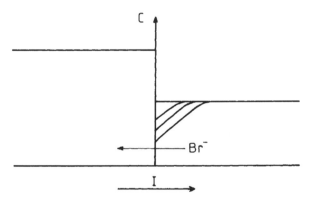

FIG. 1.   Schematic representation of the concentration profile of Br⁻ ion at different times after applying a constant current across the interface.   The bottom figure shows the effect of current reversal.

In this case, the interfacial concentration of $HTMA^+$ should decrease with time of electrolysis and the interfacial tension reaches a constant value since neither K^+ nor Br^- is surface active.

In 1968, Gavach et al. [68] showed for the first time that the interface between two immiscible electrolytes solutions (ITIES) was polarizable when the salt dissolved in the organic polar solvent was hydrophobic and that in the aqueous phase was hydrophilic. The experiments of Gavach et al. were carried out with the system alkyl (C_{12}, C_{14}, C_{16}) trimethylammonium picrate (R^+Pi^-) in nitrobenzene and with KCl in water using the following cell [68]:

$$\text{Ag} \quad | \quad \text{AgCl} \quad | \quad \underset{(w)}{R^+Cl^-} \quad | \quad \underset{(oil)}{R^+Pi^-} \quad | \quad \underset{(w)}{K^+Cl^-} \quad | \quad \text{AgCl} \quad | \quad \text{Ag}$$

These authors showed that the interface between two electrolyte solutions having a common organic cation was not polarizable, and it was therefore possible to study polarization curves for the water-nitrobenzene system by applying a constant potential difference between the two silver electrodes of the preceding cell.

Guastalla [68–73] pursued further the investigation of the polarizability of the ITIES studying what he called the hysteresis and negative conductance of the interface. A four-electrode system was used comprising two reference electrodes and two feeder electrodes as follows:

$$\text{AG} \quad | \quad \text{AgCl} \quad | \quad R^+Cl^- \quad | \quad R^+Pi^- \quad || \quad K^+Cl^- \quad | \quad \text{AgCl} \quad | \quad \text{Ag} \quad \text{Feeder}$$
$$\text{Ag} \quad | \quad \text{AgCl} \quad | \quad R^+Cl^- \quad \quad \quad KCl/AgCl/Ag \quad \text{Reference}$$

Guastalla's experimental approach was to apply a triangular voltage wave between the two feeder electrodes and to record the resulting measured potential-current relationship. These elementary studies of polarization phenomena at the liquid-liquid interface were the starting point for the use of modern electrochemical methods in the investigation of the ITIES by chronopotentiommetry (Gavach [152]), ac impedance (Gavach [155]), polarography (Koryta [178]), and cyclic voltammetry (Samec [182])(cf. Sec. VI).

III. ENERGY OF TRANSFER AND INTERFACIAL POTENTIALS

A. Introduction

The purpose of any thermodynamic treatment is to relate measurable quantities (intensive or extensive variables) in order to define the physical characteristics of the system studied. In the electrochemical system composed of two immiscible electrolytes solutions in contact, the measurable quantities are:

1. Energy of solvation of salts measured by thermochemical methods, or solubility data.

2. Distribution coefficient of salts measured by titration or other suitable analytical techniques

3. Electrical potentials measured using electrochemical concentration cells or by other potential measurements, for instance, the vibrating condenser method

The aim of this chapter is to show the relationships between these quantities and the current ideas on the possible physical pictures of the ITIES.

Consider two immiscible solvents, α and β, both containing ions of species i. Unless the system is at thermodynamic equilibrium, the work for transferring an ion i from the bulk of solvent α to the bulk of solvent β will be equal to the difference of the electrochemical potential of the ion in each phase:

$$\Delta G_{transfer}^{\alpha \to \beta} = \tilde{\mu}_i^{\beta} - \tilde{\mu}_i^{\alpha} \tag{4}$$

In order to consider an electrochemical system, it is usual to divide the electrochemical potential of the ion into a chemical and an electrical part, although Gibbs [74] and Guggenheim [75] pointed out that this splitting has no meaningful physical significance and is purely arbitrary. In this case, the change in the interaction of the ion with its neighboring atoms and molecules will be called the chemical energy of transfer, whereas the part of the work arising from the transfer of an electrical charge from the potential ϕ^α to the potential ϕ^β will be called electrical energy of transfer.

The Gibbs energy of transfer of the ion i, $\Delta G_{t,i}^{\alpha \to \beta}$ can then be written as

$$\Delta G_{t,i}^{\alpha \to \beta} = \tilde{\mu}_i^\beta - \tilde{\mu}_i^\alpha = (\mu_i^\beta - \mu_i^\alpha) + [z_i F(\phi^\beta - \phi^\alpha)] \tag{5}$$

where ϕ^α and ϕ^β are the electrical potentials in the interior of the α and β phases, called inner potential, and their difference $(\phi^\beta - \phi^\alpha)$ is called the Galvani potential difference.

B.   Ionic Standard Gibbs Energy of Transfer

To be able to express the chemical potential of an ion i in a solution, it is necessary to define for this phase a reference state (e.g., infinite dilution) and a standard state. In this way it is possible to write

$$\mu_i = \mu_i^o + RT \ln a_i \tag{6}$$

where μ_i^o is the standard chemical potential and a the activity. There-fore, if we consider the equilibrium for a salt MX,

$$M^+ + X^- \quad \text{solvated in } \alpha \text{ and} \quad \underset{\leftarrow}{\rightarrow} \quad M^+ + X^- \quad \text{solvated in } \beta \text{ and}$$
$$\text{totally dissociated} \qquad\qquad \text{totally dissociated}$$

The standard Gibbs energy of this equilibrium, which represents the standard Gibbs energy of transfer of the salt MX from α to β $\Delta G_{t,MX}^{o,\alpha \to \beta}$, is given by

$$\Delta G_{t,MX}^{o,\alpha \to \beta} = (\mu_{M^+}^{o\beta} + \mu_{X^-}^{o\beta}) - (\mu_{M^+}^{o\alpha} + \mu_{X^-}^{o\alpha}) \tag{7}$$

$\Delta G_{t,MX}^{o,\alpha \to \beta}$ is a thermodynamically measurable quantity which represents the difference in the energy of solvation or dissolution of the salt MX between

α and β. It is conceptually useful to consider it as the sum of the standard Gibbs energy of transfer of the cation $\Delta G_{t,M+}^{o,\alpha\to\beta}$ and the anion $\Delta G_{t,X-}^{o,\alpha\to\beta}$,

$$\Delta G_{t,MX}^{o,\alpha\to\beta} = \Delta G_{t,M^+}^{o,\alpha\to\beta} + \Delta G_{t,X^-}^{o,\alpha\to\beta} \tag{8}$$

with the ionic standard Gibbs energy of the ion i defined by

$$\Delta G_{t,i}^{o,\alpha\to\beta} = \mu_i^{o\beta} - \mu_i^{o\alpha} \tag{9}$$

It is important to notice that in Eq. (6) the influence of the solvent on species i is included in the standard chemical potential μ_i^o. Another approach summarized in this series by Bauer and Breant [76] is to choose a common standard state in a particular solvent (e.g, water) and to include the influence of the solvent in the activity coefficient by introducing a transfer activity coefficient. However, this alternative treatment of liquid-liquid equilibrium thermodynamics will not be treated here, as there are no advantages in the use of this formalism for the ITIES.

Having measured a full range of ΔG_t^o values for different salts, for example from solubility data, a scale of single ionic standard Gibbs energy of transfer can be established only by introducing an extra thermodynamic assumption. Many different assumptions have been proposed and can be classified into six categories [77]:

1. Assumptions based on the constant solvation energy of one ionic species in different solvents, e.g.,

   $$\Delta G_{transfer}(Rb^+) = 0 \quad \text{Pleskov [78]}$$

2. Assumptions based on the negligible difference of solvation energy between an ion and its parent neutral molecule, for those cases where the measurement is accessible, e.g.,

   $$\Delta G_{solvation}(ferrocene) = \Delta G_{solvation}(ferricenium^+) \quad \text{Strehlow [79]}$$

   $$\Delta G_{solvation}(I_2) = \Delta G_{solvation}(I_3^-) \quad \text{Alexander [80]}$$

3. Assumptions based on the negligible difference of solvation energy
    between a cation and an anion, e.g.,

$$\Delta G_{solvation}(TPAs^+) = \Delta G_{solvation}(TPB^-) = \frac{1}{2}\Delta G_{solvation}(TPAs^+TPB^-)$$

Grunwald [81]

4. Assumptions based on the negligible liquid-junction potential of the
    concentration cell, e.g.,

Ag		$AgNO_3$		$AgNO_3$		Ag
		0.01 M		0.01 M		
		S_1		S_2		

$$\Delta E = \frac{1}{F}\, \Delta G_{transfer}(Ag^+) \qquad \text{Parker [82]}$$

5. Assumptions based on the constant solvation energy of a large
    transition state for a one-step SN2 reaction (cf. Alexander and
    Parker [80]).

6. Assumptions based on the measurement of the real ionic chemical
    potential α defined by $\alpha = \mu + zF\chi$, where χ is the surface potential
    of the solvent (cf. Sec. III.E.1) (Parsons [83]).

Despite the impossibility of testing the different assumptions and the
uncertainty that an assumption is reasonable, the Grunwald assumption
has been widely used and appears to be a convenient tool for the electrochemical investigation of the ITIES. This asumption is based on the
fact that tetraphenylarsonium and tetraphenylborate are symmetrical
species of much the same size and shape where the charge is "buried"
under the phenyl groups. Therefore, their energies of solvation are
likely to be equal if the difference of solvent orientation about an anion
and a cation is small.

The Gibbs energy of transfer as described above refers to the transfer from the pure solvent α to the pure solvent β. It is therefore different from the Gibbs energy of partition, which refers to the transfer
from the solvent α saturated with β to the solvent β saturated with α.

The Gibbs energy of partition of a salt MX between α and β is defined as the Gibbs energy for the equilibrium:

$M^+ + X^-$ solvated in α saturated $\rightleftarrows$ $M^+ + X^-$ solvated in β saturated with β and totally dissociated   with α and totally dissociated

and therefore the standard Gibbs energy of partition is given by

$$\Delta G^{o,\alpha\to\beta}_{P,MX} = (\mu^{o\beta sat}_{M^+} + \mu^{o\beta sat}_{X^-}) - (\mu^{o\alpha sat}_{M^+} + \mu^{o\alpha sat}_{X^-}) = -RT \ln P^{\alpha\beta}_{MX} \tag{10}$$

with

$$P^{\alpha\beta}_{MX} = \frac{a^{\beta sat}_{M^+} a^{\beta sat}_{X^-}}{a^{\alpha sat}_{M^+} a^{\alpha sat}_{X^-}} \tag{11}$$

This value can be measured experimentally by titration of MX in each phase and by extrapolation to zero concentration. Care has to be taken in the distinction between the partition coefficient $P^{\alpha\beta}_{MX}$ above and the distribution coefficient $K^{\alpha\beta}_{MX}$:

$$K^{\alpha\beta}_{MX} = \frac{a^{\beta sat}_{\pm MX}}{a^{\alpha sat}_{\pm MX}} = \sqrt{P^{\alpha\beta}_{MX}} = \sqrt{K^{\alpha\beta}_{M^+} K^{\alpha\beta}_{X^-}} \tag{12}$$

where $K_i^{\alpha\beta}$ is the distribution coefficient of the ion i defined by

$$K_i^{\alpha\beta} = \exp\left(\frac{-\Delta G^{o,\alpha\to\beta}_{P,i}}{RT}\right) \tag{13}$$

It should be stressed that the possibility of ion pair formation in either of the phases is not taken into account if the partition coefficient as described above is expressed only in terms of concentration ratios. In this respect, the partition coefficient is different from the extraction coefficient used in phase transfer catalysis, defined as the ratio of the total concentration of MX in β over the total concentration of MX in α.

Using the same assumption as for the Gibbs energy of transfer (i.e., $\Delta G_{P,TPAs^+}^{o,\alpha\rightarrow\beta} = \Delta G_{P,TPB^-}^{o,\alpha\rightarrow\beta}$), it is also possible to define a scale of ionic standard Gibbs energy of partition. In the case of solvents of low miscibility such as water-nitrobenzene or water/1,2-dichloroethane, the ionic standard Gibbs energy of transfer is equal to the ionic standard Gibbs energy of partition showing that ions are not hydrated in the organic phase (see Table 1). This is, however, not valid for strongly hydrated ions such as Li^+ and F^-, for which Abraham and Liszi [84] have shown that their standard Gibbs energy of transfer is bigger than that of partition. Although data are available only for Li^+ and F^-, this difference between partition and transfer energy is likely to occur for ions such as Mg^{2+}, Ca^{2+}, and SO_4^{2-}.

Table 1 gives a compilation of values of standard Gibbs energy for ion transfer or partition for the water-nitrobenzene and water/1,2-dichloroethane systems all calculated with the Grunwald's assumption. More values can be found in a recent data survey of Gibbs energy of ion transfer for 57 different solvents published by IUPAC [85].

The amount of data for ionic standard Gibbs energy of transfer or partition is still rather limited. On the other hand, standard free energies of solvation of a wide range of ionic species have been measured for very usual solvents such as water or methanol. Hence the possibility of calculating standard Gibbs energy of solvation of an ion in an organic solvent using a simple model would permit prediction of the Gibbs energy of transfer from water to this solvent.

During the recent years, the model of ionic solvation of Abraham and Liszi has been used for this type of prediction [84,86–90]. Based on the observation that organic solvents are not very structured, a one-layer-continuum model has been successfully employed.

These authors first assumed that the standard Gibbs energy of solvation ΔG_s^o can be split into an electrical contribution, ΔG_e^o, and a neutral contribution ΔG_n^o defined as the Gibbs energy of solvation of a nonpolar gaseous solute of the same size as the ion considered, i.e.,

$$\Delta G_s^o = \Delta G_n^o + \Delta G_e^o \tag{14}$$

The electrical term was calculated using a one-layer solvation model in which an ion of crystallographic radius a and dielectric constant unity is surrounded by a layer of solvent of thickness $b - a$ and dielectric constant ε_1, and immersed in a bulk solvent of dielectric constant ε_0. The Gibbs energy of this simple electrostatic model is given by

$$\Delta G_e^o = \frac{N(ze)^2}{8\pi\varepsilon_0} \left[\left(\frac{1}{\varepsilon_1} - 1 \right) \left(\frac{1}{a} - \frac{1}{b} \right) + \left(\frac{1}{\varepsilon_0} - 1 \right) \frac{1}{b} \right] \tag{15}$$

The thickness of the solvent layer $b - a$ was taken as equal to the radius of the solvent molecule calculated from the bulk molar solvent volume using the Stearn-Eyring formula $r^3 = \overline{V}/8N$, and the dielectric constant ε_1 of this solvent layer is taken equal to 2 for all organic solvents. The neutral term was expressed as a first-order polynomial of the ionic radius by

$$\Delta G_n^o = ma + c \tag{16}$$

where m and c are constants of the solvent. Despite the great oversimplification of this model, the values calculated using Eqs. (14), (15), and (16) show a very reasonable agreement with the measured values. It can therefore be concluded that the Abraham and Liszi model is a useful tool for the prediction of Gibbs energies of solvation in organic solvents when no other data are available.

C. Galvani Potential Differences

The Galvani potential difference $\phi_\beta - \phi_\alpha$ between two phases at equilibrium is not a directly accessible quantity, but nevertheless, can be calculated from ionic Gibbs energy of partition when the two solvents α and β are at equilibrium. Thermodynamic equilibrium is established if

$$\tilde{\mu}_i{}^{\alpha} = \tilde{\mu}_i{}^{\beta} \tag{17}$$

and therefore by substitution of Eqs. (5), (6), and (9) into Eq. (17) the Galvani potential difference is given by

$$\Delta_\alpha{}^\beta \phi = \phi_\beta - \phi_\alpha = \frac{-1}{z_i F}\left(\Delta G^{o,\,\alpha\to\beta}_{p,i} + RT\,\ln\frac{a_i{}^\beta}{a_i{}^\alpha}\right) \tag{18}$$

This expression can also be written using limiting ionic distribution coefficient $K_i{}^{\alpha\beta}$ defined by

$$K_i{}^{\alpha\beta} = \exp\left(\frac{-\Delta G^{o,\,\alpha\to\beta}_{p,i}}{RT}\right) \tag{19}$$

and in this case the Galvani potential difference is given by

$$\Delta_\alpha{}^\beta \phi = \frac{RT}{z_i F}\,\ln\left(K_i{}^{\alpha\beta}\frac{a_i{}^\alpha}{a_i{}^\beta}\right) \tag{20}$$

By analogy with the Nernst equation, it is also possible to define for each ionic species a standard Galvani potential difference $\Delta_\alpha^\beta\phi_i^o$ given by

$$\Delta_\alpha{}^\beta \phi_i^o = \frac{-1}{z_i F}\,\Delta G^{o,\,\alpha\to\beta}_{p,i} = \frac{RT}{z_i F}\,\ln K_i{}^{\alpha\beta} \tag{21}$$

Equation (18) can then be written

$$\Delta_\alpha{}^\beta \phi = \Delta_\alpha{}^\beta \phi_i^o + \frac{RT}{z_i F}\,\ln\frac{a_i{}^\alpha}{a_i{}^\beta} \tag{22}$$

The values of the standard Galvani potential differences are also given in Table 1 for the systems water-nitrobenzene and water/1,2-dichloro-ethane. It should be stressed that standard Galvani potentials differ from the standard electrode potential, in that the former refers to an extra thermodynamic convention as the one proposed by Grunwald, whereas the latter assumes that the free energy of the reaction $H^+ + e^- \rightleftarrows \frac{1}{2}H_2$ is zero.

1. Distribution Potentials. Single Salt at Partition

In the case of a single monovalent salt at partition equilibrium, the Galvani potential difference can be calculated either from the electrochemical equilibrium of the cation or that of the anion given by Eq. (17):

$$\Delta_\alpha^\beta \phi = \Delta_\alpha^\beta \phi_+^o + \frac{RT}{F} \ln \frac{a_+^\alpha}{a_+^\beta} = \Delta_\alpha^\beta \phi_-^o + \frac{RT}{F} \ln \frac{a_-^\beta}{a_-^\alpha} \qquad (23)$$

This expression can be simplified using the limiting distribution coefficient of the ion defined earlier in Eq. (19) and it can be shown that

$$\Delta_\alpha^\beta \phi = \frac{RT}{2F} \ln \frac{K_+^{\alpha\beta}}{K_-^{\alpha\beta}} + \frac{RT}{2F} \ln \frac{\gamma_+^\alpha \gamma_-^\beta}{\gamma_+^\beta \gamma_-^\alpha} \qquad (24)$$

For dilute solutions the second term of Eq. (24) is negligible, and therefore

$$\Delta_\alpha^\beta \phi = \frac{RT}{2F} \ln \frac{K_+^{\alpha\beta}}{K_-^{\alpha\beta}} = \frac{1}{2}\left(\Delta_\alpha^\beta \phi_+^o + \Delta_\alpha^\beta \phi_-^o\right) \qquad (25)$$

This expression was first obtained by Karpfen and Randles [30], who named the Galvani potential difference in the case of a single salt at partition equilibrium, distribution potentials. The distribution potential of a salt MX is a constant that depends only at low salt concentration on the solvents α and β.

For relatively concentrated solutions, it was found experimentally by Karpfen and Randles [30], and later by Koczorowski [105] and also by Boguslavsky [106], that the term containing the activity coefficients was also close to zero. This results from the cross symmetry of this expression but should not be valid when significant ion pair formation takes place.

Although it is not possible to measure Galvani potential differences, these values can be calculated if reasonable assumptions are made regarding individual ionic activities. In dilute solutions, a common assumption is to consider that $a_+ = a_- = a_\pm$ for a 1:1 electrolyte. The two principal techniques to estimate distribution potentials are described below.

The first is a direct electrochemical method used by Karpfen and Randles [30] at the water-methyl isobutyl ketone and water-nitrobenzene interfaces, which consists in the measurement of the potential ΔE of the cell

$$\text{Hg} \mid \text{HgCl} \mid \begin{array}{c} \text{saturated} \\ \text{KCl} \end{array} \mid \begin{array}{c} \text{MX in} \\ \text{water} \end{array} \parallel \begin{array}{c} \text{MX in} \\ \text{oil} \end{array} \mid \begin{array}{c} \text{TEA}^+\text{Pi}^- \\ \text{in oil} \end{array} \mid \begin{array}{c} \text{TEA}^+\text{Pi}^- \text{ in} \\ \text{water} \end{array} \mid \text{HgPi} \mid \text{Hg}$$

$$\Delta E_8 \quad \Delta E_7 \qquad\quad \Delta E_6 \qquad \Delta E_5 \quad\ \Delta E_4 \qquad \Delta E_3 \qquad\qquad \Delta E_2 \quad \Delta E_1$$

referred to the potential ΔE_{ref} of the cell

$$\text{Hg} \mid \text{HgCl} \mid \begin{array}{c} \text{saturated} \\ \text{KCl} \end{array} \mid \begin{array}{c} \text{TEA}^+\text{Pi}^- \text{ in} \\ \text{water} \end{array} \parallel \begin{array}{c} \text{TEA}^+\text{Pi}^- \\ \text{in oil} \end{array} \mid \begin{array}{c} \text{TEA}^+\text{Pi}^- \\ \text{in oil} \end{array} \mid \begin{array}{c} \text{TEA}^+\text{Pi}^- \\ \text{in water} \end{array} \mid \text{HgPi} \mid \text{Hg}$$

$$\Delta E_8' \quad \Delta E_7' \qquad\quad \Delta E_6' \qquad\quad \Delta E_5' = \Delta E_4' \quad\ \Delta E_4' \qquad\ \Delta E_3' \qquad\quad \Delta E_2' \quad \Delta E_1'$$

From these two measurements, the distribution potential ΔE_5 is given by

$$\Delta E_5 = \Delta E - \Delta E_{ref} + \Delta E_6' - \Delta E_4 \simeq \Delta E - \Delta E_{ref} \tag{26}$$

$\Delta E_6'$ and ΔE_4 are diffusion potentials (see Sec. III.D) and are assumed to be very small since the ions TEA^+ and Pi^- have about the same ionic mobility in oil and the ions K^+ and Cl^- the ionic mobility in water. This method was recently used by Koczorowski et al. [107], who showed, furthermore, that in the case of the water-nitrobenzene and water/1,2-dichloroethane interfaces, the distribution potential ΔE_3 was close to zero since the two ions TEA^+ and Pi^- have about the same Gibbs energy of transfer (see Table 1), and therefore the distribution potential ΔE_5 was taken as equal to the measured potential ΔE_0.

The second approach to estimate distribution potentials is an indirect method first proposed by Koczorowski and Minc [108,109] in 1963 and since then widely used by Boguslavsky et al. [110–113]. It consists in the measurement of the compensation potential of the cell:

$$\text{M} \mid \begin{array}{c} \text{vibrating} \\ \text{plate} \end{array} \mid \begin{array}{c} \text{inert} \\ \text{gas} \end{array} \mid \begin{array}{c} \text{MX in} \\ \text{oil} \end{array} \mid \begin{array}{c} \text{MX in} \\ \text{water} \end{array} \mid \begin{array}{c} \text{saturated} \\ \text{KCl} \end{array} \mid \text{HgCl} \mid \text{Hg} \mid \text{M}$$

They argued that if the surface potential of the oil phase (see Sec. III. E) is independent of the concentration of MX, the change in the measured compensation potential when the salt concentration is varied represents the variation of the distribution potential.

2. Potential-Determining Ion

For a given system of two immiscible electrolyte solutions, the number of variables such as salt concentration fixing the Galvani potential difference between the two phases represents the degrees of freedom of the system, also called its variance. This is given by the difference between the sum of the number of intensive properties (T, P, μ_i, $\cdots$) for each phase and the number of restrictive conditions between them. In the case of a single salt at partition equilibrium treated above, the number of intensive variables for each phase is equal to five (T, P, μ_{M^+}, μ_{X^-}, μ_S), where S is the solvent and the restrictive conditions are the thermal and hydrostatic equilibrium, one Gibbs-Duhem relationship for each phase, one electroneutrality condition, and the two ionic transfer equilibrium $M^{+\alpha} \rightleftarrows M^{+\beta}$ and $X^{-\alpha} \rightleftarrows X^{-\beta}$. Therefore, the variance of this system is equal to $(5 + 5) - 7 = 3$ and is reduced to 1 at constant temperature and pressure.

A variance equal to 1 indicates that intensive properties such as the Galvani potential difference will be determined uniquely by a single intensive variable such as the concentration of MX. However, it is interesting to note that although the variance is not equal to zero, the distribution potential is a constant at least in the range of dilute solutions.

In systems containing more then two ions, the variance is greater than or equal to 2, and hence the Galvani potential difference will be determined if at least one ion can partition between the two phases. In this case, this ion will be called the potential-determining ion and the Galvani potential difference will be expressed from the equality of the electro-chemical potential of this ion given by Eq. (18), (20), or (22). But if no ions can partition, the interface will be blocked or polarizable and the Galvani potential difference is then undefined. The case of a polarizable interface is discussed later.

The Galvani potential difference for systems with a potential-determining ion was discussed by Melroy and Buck [114], who studied the simplest case of two salts: M^+X^-, mainly water soluble, and M^+Y^-, mainly oil soluble. The variance for such a system is equal to 1 at constant temperature and pressure, and therefore the Galvani potential differense is a function of one intensive variable of the system (e.g., the concentration of MX or the concentration of MY).

Melroy and Buck showed that for this system, where M is the potential-determing ion, the Galvani potential difference has a Nernstian response when the concentration of M^+ is varied within what these authors called a "potential window." Figure 2 illustrates the limit of validity of Nernstian response and the size of the potential window for the following values of the limiting ionic distribution coefficients: $K_{Y^-}^{o,w} = 10^6$, $K_{M^+}^{o,w} = 10^2$, and $K_{X^-}^{o,w} = 10^{-2}$.

It can be seen from Fig. 2 that the Nernstian response fails when the concentration of one salt is much bigger than that of the other salt in the other phase. In these extreme cases, the Galvani potential differences approach the distribution potential of the pure salts (i.e., +118 mV for MX and −118 mV for MY for the limiting ionic distribution coefficient given above and with the activity coefficients assumed to be equal to 1). The domain of validity of Nernstian response of the Galvani potential difference has important practical consequences in an understanding of the behavior of ion-selective electrodes. More complicated systems with potential-determining ions, where an arbitrary number of salts and the formation of ion pairs and complexes are taken into account, were analyzed by Hung [115,116].

3. Donnan Equilibrium

Donnan potential differences are defined as Galvani potential differences in the special case of two electrolyte solutions in equilibrium, where the phase boundary is not "permeable" to at least one ion, which results in an unequal distribution of "diffusible" ions [117]. In the case of liquid-liquid interfaces where no membrane is present, a Donnan equilibrium is only a limiting case of partition equilibrium, where one ion has a very

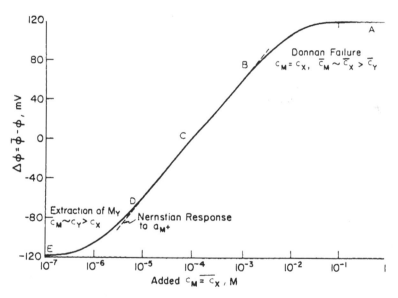

FIG. 2. An example of the potential "window." Computation for two salts: MX, mainly water soluble and MY, mainly oil soluble. Concentration of MY = 10^{-2} M; single-ion partition coefficients $K_Y = 10^6$, $K_M = 10^2$, and $K_X = 10^{-2}$. The computed potential window is 336 mV. c_M and c_X are added concentrations. (Reprinted from Ref. 114 with permission. Copyright Elsevier Science Publishers B.V., Amsterdam.)

different Gibbs energy of transfer from the other ions, so that it can be
assumed that this particular ion does not "permeate" the interface and
is present in one phase only.

Consider a sample system as shown in Fig. 3. When the system is
at thermodynamic equilibrium the Galvani potential difference, called in
this case the Donnan potential difference, can be calculated from the
equality of the electrochemical potential of M^+ and X^- between the two
phases. Therefore, we have

$$\Delta_\alpha^\beta \phi = \Delta_\alpha^\beta \phi^\circ_{M^+} + \frac{RT}{F} \ln \frac{a_{M^+}^{(\alpha)}}{a_{M^+}^{(\beta)}} \tag{27}$$

and

$$\Delta_\alpha^\beta \phi = \Delta_\alpha^\beta \phi^\circ_{X^-} + \frac{RT}{F} \ln \frac{a_{X^-}^{(\beta)}}{a_{X^-}^{(\alpha)}} \tag{28}$$

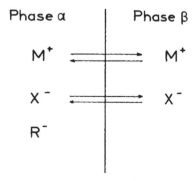

FIG. 3. Donnan equilibrium between two immiscible phases, α and β.
The ion R^- is present in one phase only whereas M^+ and X^- can partition
between the two phases.

or again by introducing the ionic distribution coefficient $K_i^{\alpha\beta}$ of M^+ and X^-, we have

$$\Delta_\alpha^\beta \phi = \frac{RT}{2F} \ln \frac{K_{M^+}^{\alpha\beta}}{K_{X^-}^{\alpha\beta}} + \frac{RT}{2F} \ln \frac{\gamma_{M^+}^{(\alpha)} \gamma_{X^-}^{(\beta)}}{\gamma_{M^+}^{(\beta)} \gamma_{X^-}^{(\alpha)}} + \frac{RT}{2F} \ln \frac{c_{M^+}^{(\alpha)} c_{X^-}^{(\beta)}}{c_{M^+}^{(\beta)} c_{X^-}^{(\alpha)}}$$

(29)

The third term of this equation can be calculated from the electroneutrality conditions for both phases:

$$c_{M^+}^{(\alpha)} = c_{X^-}^{(\alpha)} + c_{R^-}^{(\alpha)}$$

(30)

and

$$c_{M^+}^{(\beta)} = c_{X^-}^{(\beta)} = c_{MX}^{(\beta)}$$

(31)

Equation (29) can then be written

$$\Delta_\alpha^\beta \phi = \frac{RT}{2F} \ln \frac{K_{M^+}^{\alpha\beta}}{K_{X^-}^{\alpha\beta}} + \frac{RT}{2F} \ln \frac{\gamma_{M^+}^{(\alpha)} \gamma_{X^-}^{(\beta)}}{\gamma_{M^+}^{(\beta)} \gamma_{X^-}^{(\alpha)}} + \frac{RT}{2F} \ln \left[1 + \frac{c_{R^-}^{(\alpha)}}{c_X^{(\alpha)}} \right]$$

(32)

We see from this equation that if $c_{R^-}^{(\alpha)} = 0$, the Donnan potential difference becomes a distribution potential difference.

On the other hand, in the important practical case where $c_{R^-} \gg c_{X^-}$, the Donnan potential difference will have a Nernstian response with respect to the concentration of X^- in the phase α with a slope of 30 mV/decade.

Donnan potential differences at ITIES were studied by Joos et al. [118,119] in 1978 for the system KBr and $HTMA^+Br^-$ in water and $HTMA^+$ Br^- in nitrobenzene ($HTMA^+$ = hexadecyltrimethylammonium). In this system where K^+ cannot permeate through the interface, they showed that Eq. (32) was verified for dilute solutions.

D. Liquid-Junction or Diffusion Potentials

Whereas Galvani potential differences were related to two phases at thermodynamic equilibrium, liquid junctions or diffusion potentials are nonequilibrium quantities arising in the contact zone between two adjoining electrolyte solutions. The origin of such potential difference is the difference in ionic mobilities and thus in diffusion coefficients of the ionic species. In the interphase, there is therefore an ionic activity gradient which causes the diffusion.

The variation of potential in the interphase σ is defined for a zero total current by

$$\overrightarrow{\text{grad }} \phi = -\frac{1}{F} \sum_i \frac{t_i}{z_i} \overrightarrow{\text{grad }} \mu_i \tag{33}$$

where t_i is the transport number of the ion i. The total liquid junction is obtained by integration over the interphase:

$$\Delta_\alpha^\beta \phi_{\text{diff}} = -\frac{1}{F} \sum_i \int_\alpha^\beta \frac{t_i}{z_i} d\mu_i \tag{34}$$

If α and β are two electrolyte solution phases having the same solvent, different assumptions regarding the transport number profile through the diffusion layers have been proposed, leading to classical liquid-junction-potential calculations. If α and β are two immiscible solutions, one possible assumption is to define the interface as a plane P of separation. In this case Eq. (34) can be written as

$$\Delta_\alpha^\beta \phi_{\text{diff}} = -\frac{1}{F} \sum_i \int_\alpha^P \frac{t_i}{z_i} d\mu_i - \frac{1}{F} \sum_i \int_P^\beta \frac{t_i}{z_i} d\mu_i \tag{35}$$

In the first term of Eq. (35) we have

$$d\mu_i = d \ln a_i^{(\alpha)} \tag{36}$$

and in the second term,

$$d\mu_i = d \ln a_i^{(\beta)} \tag{37}$$

Therefore, by assuming the existence of a plane of separation, the liquid junction at the ITIES can be calculated using the classical liquid-junction approximations [120].

E. Volta and Surface Potentials

1. Surface Potential of Molecular Liquids

At the surface of a condensed molecular liquid, solvent molecules are the subject of orienting forces as a result of anisotropic surface. Polar or polarizable solvent molecules may then be preferentially oriented in the surface region, forming what is usually called the surface dipole layer. The potential drop associated with this layer is the surface potential χ and is positive when the positive end of the oriented dipoles points toward the bulk of the phase.

The surface potential may be expressed by the Helmholtz equation:

$$\chi = \sum \frac{N_i P_{i\perp}}{\varepsilon} \tag{38}$$

where i is the number of different solvents in the liquid, N the number of molecules of solvent i per unit area, $P_\perp$ the component normal to the interface of the dipole of the molecule i, and ε the local permittivity. In the case of an electrolyte solution, oppositely charged ionic components can penetrate the surface region differently, forming surface ion pairs that can also be oriented and contribute to the surface potential, as shown in Fig. 4.

A recent study of the surface potential of solutions of tetraalkyl-ammonium salts carried out by Bennes and Conway [121] showed that the orientation of ion pairs at the air-water interface contributes to the surface potential and that interfacial solvation models can be derived from surface potential measurements.

2. *Interface Between Two Molecular Liquids*

When two immiscible liquids phases α and β are brought in contact, an interfacial phase is created. This interphase can be defined as the interfacial region where the properties vary between those of the bulk. If the two phases are ideally immiscible, the interphase can be represented by two surface regions separated by an interface, as shown in Fig. 5.

If the two liquids are not ideally immiscible, as is the case between immiscible polar solvents, there is then an interfacial region where solvent mixing occurs. The interphase can then be represented by two surface regions that intermix as shown in Fig. 6. At a newly created interphase between polar liquids, the solvent molecules become preferentially oriented by the short-range anisotropic forces between them. As

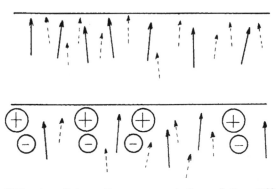

FIG. 4. Schematic representation of the different dipolar contributions to the surface potential for a mixture of two liquids (top) and two liquids with a salt that can be adsorbed at the interface forming ion pairs (bottom).

a result of this preferential orientation, a potential drop g_α^β (dip) is therefore established.

For the purpose of modeling the interfacial region, this dipolar contribution to the Galvani potential difference can be expressed by [131]

$$g_\alpha^\beta(\text{dip}) = g_\beta^{(\alpha)}(\text{dip}) - g_\alpha^{(\beta)}(\text{dip}) \tag{39}$$

where $g_\beta^{(\alpha)}$ and $g_\alpha^{(\beta)}$ are the dipolar contribution of β and α when the interphase is formed. The potential drop contribution $g_i^{(j)}$ of phase i when brought in contact with phase j is related to the surface potential χ_i of the phase i by

$$g_i^{(j)} = \chi_i + \delta\chi_i^{(j)} \tag{40}$$

where $\delta\chi_i^{(j)}$ represents the deviation from the surface potential caused by the presence of the phase j at the interface.

Therefore, the potential drop $g_\alpha^{(\beta)}(\text{dip})$ is given by

$$g_\alpha^\beta(\text{dip}) = (\chi_\beta - \chi_\alpha) + \delta\chi_\beta^{(\alpha)} - \delta\chi_\alpha^{(\beta)} \tag{41}$$

or

$$g_\alpha^\beta(\text{dip}) = \Delta_\alpha^\beta\chi + \delta\Delta_\alpha^\beta\chi \tag{42}$$

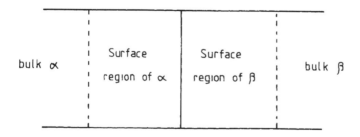

FIG. 5. The interfacial region between phases α and β shown as two immiscible surface regions with a plane of separation between them.

It can be seen, then, that $g_\alpha^\beta(\text{dip})$ cannot be calculated even if the surface potential for each liquid is known.

3.   *Interface Between Two Electrolyte Solutions*

At the interface between two immiscible electrolyte solutions (ITIES), the Galvani potential difference between the two phases α and β, $\Delta_\alpha^\beta \phi$, can be divided into a dipolar contribution $g_\alpha^\beta(\text{dip})$ arising only from solvent molecule orientation and an ionic contribution $g_\alpha^\beta(\text{ion})$ associated with the ionic charges in the interphase, i.e.,

$$\Delta_\alpha^\beta \phi = g_\alpha^\beta(\text{ion}) + g_\alpha^\beta(\text{dip}) \tag{43}$$

This separation into dipolar and ionic contribution is conceptually similar to the separation of the Galvani potential difference into a Volta $\Delta_\alpha^\beta \psi$ and a surface $\Delta_\alpha^\beta \chi$ potential difference, illustrated in Fig. 7 and given by

$$\Delta_\alpha^\beta \phi = \Delta_\alpha^\beta \psi + \Delta_\alpha^\beta \chi \tag{44}$$

However, there is a significant difference between Eqs. (43) and (44), as the dipolar contribution $g_\alpha^\beta(\text{dip})$ is not equal to the difference of the surface potential $\Delta_\alpha^\beta \chi$. As in Sec. III.E.2, these two values are related by Eq. (42). Consequently, the dipolar contribution $g_\alpha^\beta(\text{dip})$ cannot be calculated even though the difference $\Delta_\alpha^\beta \chi$ can be measured.

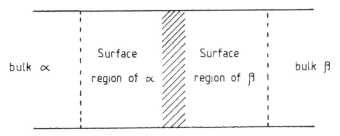

FIG. 6.   Same as Fig. 5, but assuming that a mixed solvent layer (shaded area) is present at the interfacial region.

Nevertheless, it will be shown in Sec. V that in the case of polarized or ideally polarized interfaces, $g_\alpha^\beta(dip)$ can be estimated from the measurement of the potential of zero charge (pzc) since in this particular case condition Eq. (45) applies.

$$g_\alpha^\beta(ion)_{pzc} = 0 \tag{45}$$

The Volta potential difference at ITIES was recently studied by Koczorowski and Zagorska [122] for a simple system consisting of a single salt MX at partition equilibrium between water and nitrobenzene. They showed that the Volta potential difference at this ITIES could be estimated by measuring the compensation potential ΔE_1 and ΔE_2 of the two cells.

| M | vibrating plate | inert gas | oil MX | water MX | saturated calomel | M | ΔE_1 |

and

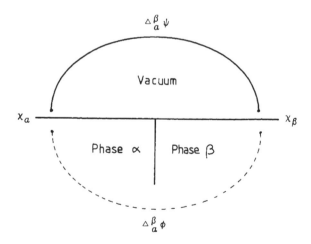

FIG. 7. Relationship between the Volta and the surface potentials of two phases α and β, and the Galvani potential difference.

| M | vibrating plate | inert gas | water saturated with oil MX | saturated calomel | M | ΔE_2 |

Indeed, if we assume the diffusion potential in the aqueous phase to be negligible, ΔE_1 is equal to

$$\Delta E_1 = \Delta_w^o \phi + \chi_o + \text{const.} \tag{46}$$

and ΔE_2 is equal to

$$\Delta E_2 = \chi_w + \text{const.} \tag{47}$$

and therefore the Volta potential difference $\Delta_w^o \psi$ can be calculated from

$$\Delta_w^o \psi = \Delta_w^o \phi - \Delta_w^o \chi = \Delta E_1 - \Delta E_2 \tag{48}$$

The values obtained for different salts are given in Table 2.

These authors also showed that by measuring the compensation potential ΔE_3 of the cell

| M | vibrating plate | inert gas | oil MX | oil TEA P_i | water TEA P_i | saturated calomel | M | ΔE_4 |

it was possible to estimate the surface potential difference $\Delta_w^o \chi$ using the following assumptions:

TABLE 2

	$\Delta\psi_w^o \pm 10$ mV	$\Delta\phi_w^o \pm 5$ mV	$\Delta_w^o \chi$
TPACl	485	375	−110
TBACl	440	320	−120
TEACl	310	200	−110
KCl	215		
NaCl	180		
NaBr	128		

Source: From Ref. 122.

1. The diffusion potentials in the water phase are negligible.

2. The diffusion potentials in the oil phase are negligible because TEA^+ and Pi^- have the same ionic mobility in nitrobenzene.

3. The distribution potential of TEA^+Pi^- is negligible since the Gibbs energy of partition of this salt is close to zero (see Table 1).

The surface potential difference $\Delta_w^o \chi$ is then given by

$$\Delta_w^o \chi = \Delta E_3 - \Delta E_2 \tag{49}$$

The values found for the different salts were more or less constant and found to be equal to

$$\Delta_w^o \chi = -105 \pm 20 \text{ mV} \tag{50}$$

It is important to notice that the compensation potential ΔE_2 and ΔE_3 are concentration dependent. As Bennes and Conway [121] had shown earlier, the variation of surface potential at the air-water interface when the salt concentration is varied is quite considerable, as shown in Fig. 8 [122]. Consequently, even if the Galvani distribution potential is constant over a wide range of concentration, this does not imply that $\Delta \psi$ and $\Delta \chi$ are also constant within this range of concentration.

IV. INTERFACIAL THERMODYNAMICS

A. Gibbs Adsorption Equation

The thermodynamic treatment of ionic systems can be performed using two different approaches. Either the salts are considered as a neutral component, in which case the treatment follows closely that for nonionic systems, or the salts are considered as ionic species, in which case electrochemical potentials are used to describe the system. Parsons [123] has shown that these two approaches are equivalent, and the second option was chosen arbitrarily for the present thermodynamic analysis of ITIES.

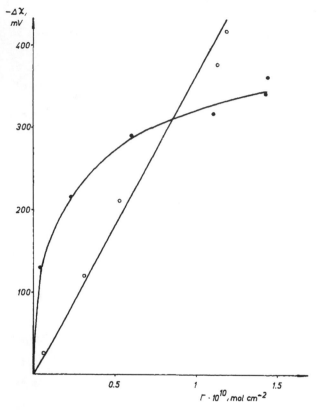

FIG. 8. Surface potential change $\Delta\chi$ as a function of surface excess Γ for solutions of TBACl in water (open circles) and in water saturated with nitrobenzene (filled circles). (Reprinted from Ref. 122 with permission. Copyright Elsevier Science Publishers B.V., Amsterdam.)

Following Guggenheim's approach, we shall consider the interfacial region between two immiscible electrolyte phases α and β as a separate interfacial phase σ.

If the phase α is composed of one solvent S_α containing m ionic species and the phase β of one solvent S_β containing p ionic species, the number of components of the interphase is m + p + 2 − c, where c is the number of common ions between the two phases. The Gibbs energy of the interphase (σ) is given by

$$dG^\sigma = -S^\sigma\,dT + V^\sigma\,dP + A d\gamma + \sum_{i=1}^{m+p-c} \tilde{\mu}_i\,dn_i^\sigma + \mu_{s_\alpha}\,dn_{s_\alpha}^\sigma$$
$$+ \mu_{s_\beta}\,dn_{s_\beta}^\sigma \tag{51}$$

where S^σ is the interfacial entropy, V^σ the volume of the interphase, γ the interfacial tension, n_i^σ the number of ions i in the interphase, and $n_{s_\alpha}^\sigma$ and $n_{s_\beta}^\sigma$ the number of molecules of solvent in the interphase.

In Eq. [51] all the intensive properties of the interphase, S, P, γ, $\tilde{\mu}_i$, and μ_s, are not independent variables. The number of independent variables is given by the variance or degree of freedom of the system defined by the sum of the numbers of intensive properties for each phase less the number of restrictive conditions between them.

In the present system, the intensive variables of the phase α are the chemical potential of the m + 1 components, T and P, and those of the phase β are p + 1 components, T and P. On the other hand, the restrictive conditions are a Gibbs-Duhem relationship for each phase (Eqs. (52) and (53)], the thermal and hydrostatic equilibrium, the electroneutrality of the system, and if there are c common ions between the two phases, c partition equilibrium conditions. The variance of this system is therefore equal to $(m + 3) + (p + 3) - (5 + c) = M + p + 1 - c$ and is reduced to $m + p - 1 - c$ at constant temperature and pressure.

When the two solvents are partially miscible, the variance remains the same, as the increase in the number of components in each phase is compensated by the same increase in the number of restrictive conditions resulting from the solvent partition equilibria. Similarly, the variance remains unaltered when incomplete dissociation of the salts is considered, as any increase of component for considering the undissociated ion pair as a neutral component is balanced by the restrictive ion pair formation equilibrium.

If the solvents are not perfectly immiscible, the Gibbs-Duhem relationships for the two bulk phases will be

$$S^\alpha \, dT - V^\alpha \, dP + n_{s_\alpha}{}^\alpha \, d\mu_{s_\alpha} + n_{s_\beta}{}^\alpha \, d\mu_{s_\beta} + \sum_{i=1}^{m} n_i{}^\alpha \, d\tilde{\mu}_i = 0 \qquad (52)$$

and

$$S^\beta \, dT - V^\beta \, dP + n_{s_\alpha}{}^\beta \, d\mu_{s_\alpha} + n_{s_\beta}{}^\beta \, d\mu_{s_\beta} + \sum_{i=1}^{p} n_i{}^\beta \, d\tilde{\mu}_i = 0 \qquad (53)$$

The analog of the Gibbs-Duhem equation for the interphase, called the Gibbs adsorption equation, is given by

$$S^\sigma \, dT - V^\sigma \, dP + A \, d\gamma + n_{s_\alpha}{}^\sigma \, d\mu_{s_\alpha} + n_{s_\beta}{}^\sigma \, d\mu_{s_\beta} + \sum_{i=1}^{m+p-c} n_i{}^\sigma \, d\tilde{\mu}_i = 0$$

$$(54)$$

At constant temperature and pressure, Eq. (54) is more conveniently expressed as

$$-d\gamma = \Gamma_{s_\alpha} \, d\mu_{s_\alpha} + \Gamma_{s_\beta} \, d\mu_{s_\beta} + \sum_{i=1}^{m+p-c} \Gamma_i \, d\tilde{\mu}_i \qquad (55)$$

where $\Gamma = n^\sigma / A$ represents surface concentrations.

Using Eqs. (52) and (53) it is possible to calculate $d\mu_{s_\alpha}$ and $d\mu_{s_\beta}$ as a function of $d\tilde{\mu}_i$ at constant temperature and pressure, and to substitute the calculated value into Eq. (55), which becomes

$$-d\gamma = \sum_{i=1}^{m+p-c} \Gamma_i^{\alpha,\beta} \, d\tilde{\mu}_i \tag{56}$$

with $\Gamma_i^{\alpha,\beta}$ the relative surface excess concentration of the ion i with respect to both solvents defined by

$$\Gamma_i^{\alpha,\beta} = \Gamma_i - n_i^\alpha \frac{\Delta_\alpha}{\Delta} - n_i^\beta \frac{\Delta_\beta}{\Delta} \tag{57}$$

where

$$\Delta = \begin{vmatrix} n_{s_\beta}^\alpha & n_{s_\alpha}^\alpha \\ n_{s_\beta}^\beta & n_{s_\alpha}^\beta \end{vmatrix} \qquad \Delta_\alpha = \begin{vmatrix} n_{s_\alpha}^\beta & \Gamma_\alpha \\ n_{s_\beta}^\beta & \Gamma_\beta \end{vmatrix} \qquad \Delta_\beta = \begin{vmatrix} n_{s_\beta}^\alpha & \Gamma_\beta \\ n_{s_\alpha}^\alpha & \Gamma_\beta \end{vmatrix} \tag{58}$$

For solvent of low mutual solubility such as water-nitrobenzene and water/1,2-dichloroethane, where $n_{s_\alpha}^\beta \ll n_{s_\alpha}^\alpha$ and $n_{s_\beta}^\alpha \ll n_{s_\beta}^\beta$, Eq. (57) is reduced to

$$\Gamma_i^{\alpha,\beta} = \Gamma_i - \frac{n_i^\alpha}{n_{s_\alpha}^\alpha} \Gamma_\alpha - \frac{n_i^\beta}{n_{s_\beta}^\beta} \Gamma_\beta \tag{59}$$

As any surface excess quantity, $\Gamma_i^{\alpha,\beta}$ is independent of the volume of the interphase v^σ.

Equation (56) is the general form of the Gibbs equation for an ITIES. Nevertheless, it is convenient to express the electrochemical potentials of the ions in terms of measurable chemical potentials of salts. In this case, the interfacial tension, at constant temperature and pressure, is given as a linear combination of the $m + p - 1 - c$ independent intensive variables

$$-d\gamma = \sum_{i=1}^{m+p-1-c} \Psi_{salt}\, d\mu_{salt} \tag{60}$$

where Ψ_{salt} is the linear combination of independent variables associated to the intensive variable $d\mu_{salt}$. Equation (60) has the same form as that for a system of neutral molecules derived previously by Hansen [124] and Defay et al. [125].

B. Ionic Adsorption and Surface Excess of Water

In the same way that the Gibbs adsorption equation for two-phase systems with neutral solutes is used to measure adsorption phenomena, the Gibbs adsorption equation for an ITIES can be used to study ionic adsorption.

Consider as an example the simple system of a salt MX at partition equilibrium between two phases α and β. It can be shown from Eq. (56) and from the electroneutrality condition of the interphase that

$$\Gamma_{M^+}^{\alpha,\beta} = \Gamma_{X^-}^{\alpha,\beta} = \Gamma_{MX}^{\alpha,\beta} \tag{61}$$

with

$$\Gamma_{MX}^{\alpha,\beta} = -\frac{1}{2RT}\left(\frac{\partial\gamma}{\partial\ln a_{\pm MX}}\right)_{T,P} \tag{62}$$

Equation (62) has been derived by Boguslavsky et al. [126,127] and by Gavach et al. [128,129]. Both authors studied the adsorption isotherm of tetraalkylammonium halides partitioned between water and nitrobenzene. The experimental results found by these two groups are very similar, and although the theoretical considerations leading to a physical model follow different pathways, the models of ionic adsorption proposed are rather similar.

Boguslavsky et al. [127] fitted their data to a Frumkin isotherm and concluded that for tetraethylammonium chloride, which is highly insoluble in nitrobenzene, there was no penetration of ionic species in the oil phase, whereas for longer alkyl chains there was penetration of the cations in the oil phase, as shown in Fig. 9. These authors further justified the proposed model from surface potential measurements.

On the other hand, Gavach et al. [129] fitted their data to an adaptation of the Gouy-Chapman theory to the ITIES (see Sec. V.F) and also suggested the possible formation of interfacial ion pairs as shown in Fig. 9.

For the same system of a single salt at partition, it is also possible to measure the adsorption of one of the solvents at the interface. Starting from Eq. (55), which for the present case reads

$$-d\gamma = \Gamma_{S_\alpha} d\mu_{S_\alpha} + \Gamma_{S_\beta} d\mu_{S_\beta} + \Gamma_{MX} d\mu_{MX} \tag{63}$$

it can be seen that

$$\Gamma_{S_\alpha}^{\beta,MX} = -\left(\frac{\partial \gamma}{\partial \mu_{S_\alpha}}\right)_{T,P} \tag{64}$$

where $\Gamma_{S_\alpha}^{\beta,MX}$ is the relative surface excess of the solvent S_α with respect to the solvent S_β and MX given by

$$\Gamma_{S_\alpha}^{\beta,MX} = \Gamma_{S_\alpha} + \Gamma_{S_\beta} \frac{\Delta_{S_\beta}}{\Delta} + \Gamma_{MX} \frac{\Delta_{MX}}{\Delta} \tag{65}$$

with

$$\Delta = \begin{vmatrix} n_{S_\beta}^\alpha & n_{MX}^\alpha \\ \\ n_{S_\beta}^\beta & n_{MX}^\beta \end{vmatrix} \quad \Delta_{S_\beta} = \begin{vmatrix} n_{S_\alpha}^\alpha & n_{MX}^\alpha \\ \\ n_{S_\alpha}^\beta & n_{MX}^\beta \end{vmatrix} \quad \Delta_{MX} = \begin{vmatrix} n_{S_\beta}^\alpha & n_{S_\alpha}^\alpha \\ \\ n_{S_\beta}^\beta & n_{S_\alpha}^\beta \end{vmatrix} \tag{66}$$

For solvents of low mutual solubilities, such as water-nitrobenzene or water/1,2-dichloroethane, where $n_{s_\alpha}{}^\alpha \gg n_{s_\alpha}{}^\beta$ and $n_{s_\beta}{}^\beta \gg n_{s_\beta}{}^\alpha$, the determinant Δ_s is reduced to $-n_{s_\alpha}{}^\alpha n_{s_\beta}{}^\beta$. Further simplifications occur if the partition coefficient of the salt is very large (i.e., $n_{MX}{}^\alpha \gg n_{MX}{}^\beta$). In this case, Eq. (65) is reduced to

$$\Gamma_{s_\alpha}{}^{\beta, MX} = \Gamma_{s_\alpha} \tag{67}$$

Girault and Schiffrin derived Eqs. (65) and (67) and measured the adsorption of water at the interface with nitrobenzene, 1,2-dichloroethane, and heptane [130]. The relative surface excesses of water obtained were expressed in terms of monolayers of water, assuming that a monolayer represents 1.72×10^{-9} mole/cm^2 and the results available are collected in Table 3.

The surface excesses of water at an ITIES were positive quantities similar to those measured at the air-water or mercury-water interfaces, but the values obtained were smaller than one monolayer for the polar organic solvents. It was proposed that at the interface between two polar solvents, intermixing occurs which may be accompanied by mixed

NITROBENZENE

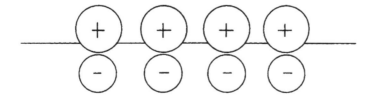

TBA Cl in WATER

FIG. 9.   Penetration into the organic phase of hydrophobic ions dissolved in water.

TABLE 3

Values of the Ion-Free Layer Thickness, δ (Expressed as Monolayers of Water, Assuming that $\delta = 1$ Corresponds to 1.72×10^{-9} Mole cm^{-2} of Water), at the Interface Between Water and Immiscible Media

	LiCl	NaCl	KCl	MgSO$_4$
Nitrobenzene	0.4	0.7	0.4	3.0
1,2-Dichloroethane	0.4	0.9	0.6	—
n-Heptane	0.9	—	1.0	—
Air-solution	0.9	1.0	1.0	2.2

Source: From Ref. 130.

solvation of the ions. In fact, as Trasatti and Parsons [131] pointed out, ions can approach the interface to an extent depending on the forces in play. In the case of inorganic salts, these forces are only long-range interactions and the ions are not specifically adsorbed.

In the case of organic salts such as TBACl, short-range forces are also present, due to dipole-induced dipole and dispersive interactions. Then, if the model proposed by Boguslavsky or Gavach of interfacial ion pairs is valid, it can be said that the organic cations are specifically adsorbed, and that in terms of the models of the mercury-solution interface commonly used, the number of equivalent adsorption sites is concentration dependent.

The change in surface excesses of water with the polarity of the organic immiscible solvent had already been observed by Aveyard and Saleem [132a,b], who studied ionic desorption (equivalent to solvent adsorption) at the interface between alkali halide electrolyte solution and a wide range of organic solvents. Using Eq. (62) they showed that the surface excesses of water decreased when the polarity of the organic solvent increased for the following solvents:

alkanes < cyclohexane < chlorooctane < benzene $\simeq$ toluene

< methylalkanoate < nonanone < decanol

They concluded that polar liquids were more able to promote interfacial water structure than apolar solvents and that consequently, interfacial ionic penetration was deeper at the interface between water and polar liquids.

C. Adsorption of Neutral Molecules

The presence in the system of m' neutral species in the phase α and p' neutral species in the phase β considered in Sec. IV.A will increment the variance of the system by $m' + p' - c'$, where c' is the number of common neutral species between the two phases. In this case the Gibbs equation (56) becomes

$$-d\gamma = \sum_{i=1}^{m+p-c} \Gamma_i^{\alpha,\beta} \, d\tilde{\mu}_i + \sum_{j=1}^{m'+p'-c'} \Gamma_j^{\alpha,\beta} \, d\mu_j \tag{68}$$

where $\Gamma_j^{\alpha,\beta}$ is the relative surface excess concentration of the molecule j with respect to both solvents defined by

$$\Gamma_j^{\alpha,\beta} = \Gamma_j - n^\alpha \frac{\Delta_\alpha}{\Delta} - n_j^\beta \frac{\Delta_\beta}{\Delta} \tag{69}$$

with Δ, Δ_α, and Δ_β being defined by Eq. (58). Adsorption of neutral molecules at liquid-liquid interfaces is a very important subject of the physical chemistry of liquid-liquid interfaces which goes beyond the scope of this review. For further reading, see, for example, Ref. 133.

V.   ELECTRIFIED ITIES

A.   Gibbs Adsorption Equation for the Electrified ITIES

When an interface between two immiscible solutions is electrified, that is, when a potential difference can be externally applied to the interface, it is convenient to express the applied potential difference as a function of the electrochemical potentials of the ionic species involved in the reference electrode equilibria. We consider as an example that the electrode in phase α is reversible to an anion A_r^- and that the one present in phase β is reversible to a cation C_r^+. It is easy to show that if the two reversible electrodes are in contact with the same metal (e.g., Cu), the applied potential difference $\Delta_\alpha^\beta E = E_\beta - E_\alpha$ is defined as $E_{+(-)}$ and given by

$$d\Delta_\alpha^\beta E = dE_{+(-)} = \frac{1}{F} d\tilde{\mu}_{C_r^+} + d\tilde{\mu}_{A_r^-} \tag{70}$$

or

$$dE_{+(-)} = \frac{1}{F} d\mu_{C_r^+ A_r^-} \tag{71}$$

The subscript $+(-)$ refers to the types of reference electrodes used in phases β and α.

Therefore, it can be seen that the applied potential difference $dE_{+(-)}$ can be introduced in the Gibbs adsorption equation (60) by replacement of the term relative to the salt $C_r A_r$. The Gibbs equation for an electrified ITIES is therefore

$$-d\gamma = \sum_{1=1}^{m+p-2-c} \Psi_{salt} \, d\mu_{salt} + Q_{C_r A_r} \, dE_{+(-)} \tag{72}$$

where $Q_{C_r A_r}$ is the extensive variable associated with the applied potential difference defined by

$$Q_{C_r A_r} = F\Psi_{C_r A_r} \tag{73}$$

From Eqs. (72) and (73),

$$-\left(\frac{\partial\gamma}{\partial E_{+(-)}}\right)_{\mu_{salt}} = Q_{C_r A_r} \tag{74}$$

which is the Lippmann equation for the electrified ITIES. As Frumkin [134] pointed out, $Q_{C_r A_r}$ is the amount of electricity flowing to the interphase when its area is increased by a unit amount. Equation (72) represents the basic equation for the thermodynamic description of an electrified ITIES and it is convenient to show how these equations can be used to treat two idealized types of interfaces:

1. Ideally unpolarized interface
2. Ideally polarized interface

A real ITIES may approach the behavior of these abstractions under certain conditions which are discussed below.

B. Ideally Unpolarizable ITIES

The characteristic of a perfectly unpolarizable ITIES is the occurrence of an unhindered exchange of ions between the two phases, and therefore ionic equilibrium for all the ionic species prevails throughout the system. Let us illustrate this case with an example approaching the conditions of ideal unpolarizability. Consider the salt KCl in water supposed totally insoluble in oil, and the salt C^+A^- in the organic phase supposed totally water insoluble, and a potential-determining ion M^+ distributed between the two phases. At constant temperature and pressure, the variance $m + p - 1 - c$ of this system is then equal to $3 + 3 - 1 - 1 = 4$.

The Gibbs adsorption equation (56) is

$$-d\gamma = \Gamma^{o,w}_{K^+} d\bar{\mu}_{K^+} + \Gamma^{o,w}_{Cl^-} d\bar{\mu}_{Cl^-} + \Gamma^{o,w}_{C^+} d\bar{\mu}_{C^+} + \Gamma^{o,w}_{A^-} d\bar{\mu}_{A^-}$$
$$+ \Gamma^{o,w}_{M^+} d\bar{\mu}_{M^+} \qquad (75)$$

Let us assume furthermore that the electrode in the aqueous phase is reversible to Cl^- and that in the oil phase is reversible to C^+. Consequently, $dE_{+(-)}$ is given by

$$dE_{+(-)} = \frac{1}{F} d\mu_{CCl} \qquad (76)$$

The introduction of $dE_{+(-)}$ in Eq. (75) leads to

$$-d\gamma = \Gamma^{o,w}_{K^+} d\mu_{KCl} + \Gamma^{o,w}_{A^-} d\mu_{CA} + \Gamma^{o,w}_{M^+} d\mu_{MCl} + Q_{CCl} dE_{+(-)} \qquad (77)$$

Using the electroneutrality condition of the interphase, it is easy to show that

$$Q_{CCl} = F\left(\Gamma_{C^+} - \Gamma_{A^-}\right) = F\left(\Gamma_{Cl^-} - \Gamma_{K^+} - \Gamma_{M^+}\right) \qquad (78)$$

Let us now consider the case where the reversible electrode in the oil phase is reversible toward M^+. In this case, Eq. (75) remains unaltered, but the introduction of the electrical term $dE_{+(-)}$ now equal to $d\mu_{MCl}/F$ leads to

$$-d\gamma = \Gamma_{K^+} d\mu_{KCl} + \Gamma_{A^-} d\mu_{CA} + \left(\Gamma_{C^+} - \Gamma_{A^-}\right) d\mu_{CCl} + Q_{MCl} dE_{+(-)} \qquad (79)$$

with

$$Q_{MCl} = F\Gamma_{M^+} = F\left(\Gamma_{Cl^-} + \Gamma_{A^-} - \Gamma_{K^+} - \Gamma_{C^+}\right) \qquad (80)$$

Equations (78) and (80) show that the quantity of electricity Q flowing to the interface when its area is increased by a unit amount is different and depends on the choice of the two reference electrodes.

As Parsons pointed out, these equations are just an alternative way of expressing a surface excess or a combination of surface excesses of ionic species [135]. In fact, the values Q_{MCl} could be measured experimentally by measuring the interfacial tension when the concentration of MCl in one phase is varied. This approach was carried out by Gros et al. for the system [99]

RBr variable $\quad \|$ RTPB 10^{-2} M
NaBr 3×10^{-2} M $\|$

and the results obtained are shown in Fig. 10 (R = tetraalkylammonium ion).

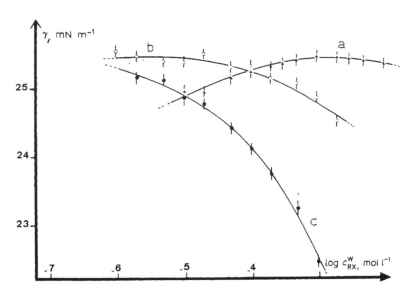

FIG. 10. Variation of the interfacial tension of a system constituted by a solution of 10^{-2} M tetraalkylammonium tetraphenylborate ($R^+R'^-$) in nitrobenzene and an aqueous solution of 3×10^{-2} M NaBr and tetraethylammonium bromide (RBr) at variable concentrations. R^+ = (a) Tetraehtylammonium; (b) tetrapropylammonium; (c) tetrabutylammonium ion. The potentials are related to the interfacial Galvani potential difference by: $-\Delta\phi = \Delta_O^W \phi + 0.120 \pm 0.005$ V. (Reprinted from Ref. 99 with permission. Copyright Elsevier Science Publishers B.V., Amsterdam.)

For ideally unpolarized interfaces, even assuming the presence of a physical barrier between the two phases, it is not possible to define the physical electrical charge located on the aqueous side of the barrier and that located in the organic side. This is due to the fact that the potential-determining ion passes freely through the interface, and therefore it is not possible to divide the surface excess of this ion into an aqueous and a nonaqueous contribution. Nevertheless, if the supporting electrolyte (i.e., the ions which do not partition) is in excess compared with the concentration of the potential-determining ion, a physical model regarding the nature of the interfacial charge can be made if the relative surface excess of this ion is neglected.

In the example treated here the assumption that $C_{KCl} \gg C_{MCl}$ and $C_{CA} \gg C_{MA}$ leads to

$$\Gamma_{K^+} + \Gamma_{M^+} - \Gamma_{Cl^-} \approx \Gamma_{K^+} - \Gamma_{Cl^-} = q_{water} \tag{81}$$

and

$$\Gamma_{C^+} + \Gamma_{M^+} - \Gamma_{A^-} \approx \Gamma_{C^+} - \Gamma_{A^-} = q_{oil} \tag{82}$$

We see from these equations that the assumption of a physical model of the interface (i.e., assuming a plane of separation), and the assumption of negligible concentration of potential determining ions allow the definition of an interfacial charge such as

$$q_{water} = -q_{oil} \tag{83}$$

In the system studied by Gros et al. [99], similar assumptions lead to

$$\Gamma_{Na^+} + \Gamma_{R^+} - \Gamma_{Br^-} \approx \Gamma_{Na^+} - \Gamma_{Br^-} = q_{water} \tag{84}$$

and

$$\Gamma_{R^+} - \Gamma_{TPB^-} \approx q_{oil} \tag{85}$$

For this system, the Gibbs equation reads

$$-d\gamma = \left(\Gamma_{R^+} - \Gamma_{TPB^-} \right) d\mu_{RBr} + \Gamma_{Na^+} d\mu_{NaBr} + \Gamma_{TPB^-} d\mu_{TPB^-} \tag{86}$$

or

$$-d\gamma = Q_{KBr} \, dE_{+(-)} + \Gamma_{Na^+} d\mu_{NaBr} + \Gamma_{TPB^-} d\mu_{TPB^-} \tag{87}$$

It can be seen, therefore, that in the particular case where the concentration of the potential determining ion is small compared to the other electrolytes, the extensive electrical variable Q_{KBr} represents the state of charge of the interface, as defined by Eqs. (84) and (85).

Furthermore, these authors have shown that for the example studied,

$$d\mu_{RBr} \simeq d\Delta_\alpha^\beta \phi \tag{88}$$

where $\Delta_\alpha^\beta \phi$ is the Galvani potential difference, and consequently by plotting the interfacial tension values measured against, $\Delta_\alpha^\beta \phi$ expressed on an arbitrary scale, the classical electrocapillary curves are obtained (see Fig. 11).

Nevertheless, the location of the physical charge, q_{water} and q_{oil}, cannot be derived from the thermodynamic analysis and depends entirely on the physical model asumed. This is discussed thoroughly in Sec. V. C.1.

C.   Ideally Polarized ITIES

The characteristic of an ideally polarized ITIES is the absence of exchange of ions between the two phases. Consequently, the number of ions that can partition is obviously equal to zero and the variance of the system is equal to $n + p - 1$ at constant temperature and pressure. The equilibrium setup at a perfectly polarizable ITIES is therefore not chemical but purely electrostatic.

Let us illustrate this case with an example and consider a hydrophobic salt CA totally insoluble in water and a hydrophilic salt KCl assumed to be totally insoluble in oil. Let us also assume that the electrode in the aqueous phase is reversible to Cl^-, and that in the oil phase is reversible to C^+. For this particular case, Eq. (72) becomes

$$-d\gamma = \Gamma_{K^+}^{\alpha,\beta} d\mu_{KCl} + \Gamma_{A^-}^{\alpha,\beta} d\mu_{CA} + Q_{CCl} dE_{+(-)} \tag{89}$$

with

$$Q_{CCl} = F\left(\Gamma_{C^+} - \Gamma_{A^-}\right) = F\left(\Gamma_{Cl^-} - \Gamma_{K^+}\right) \tag{90}$$

By definition, Q_{CCl} is again the quantity of electricity flowing to the interphase when its area is increased by a unit amount, but it can easily

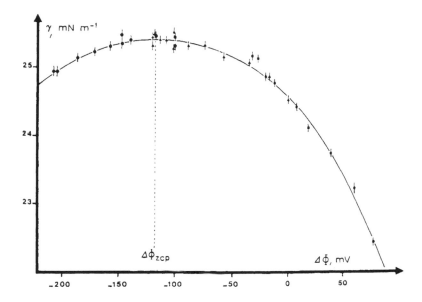

FIG. 11. Electrocapillary curve: variation of the interfacial tension with the Galvani potential difference $\Delta\phi$. Filled circle, square, and triangle, experimental values and solid line, calculated values obtained by integration of charge derived from the Gouy-Chapman theory. (Reprinted from Ref. 99 with permission. Copyright Elsevier Science Publishers B.V., Amsterdam.)

be shown that whatever the reference electrodes chosen, the electrical variable associated to $dE_{+(-)}$ or $dE_{-(+)}$ is a constant of the system studied.

1. Electrical Charge

Contrary to the ideally unpolarizable interface, the electrical variable is uniquely defined for ideally polarizable interfaces and in what follows will be called Q. Therefore, the Lippmann equation, given by

$$-\left(\frac{\partial\gamma}{\partial E}\right)_{\mu_{salt}T,P} = Q \tag{91}$$

is a characteristic of the interface and not a variable associated with a particular salt. This equation is, furthermore, the expression of electro-

capillarity phenomena showing the dependence of the interfacial tension
on the applied potential difference.

Q is an experimentally accessible quantity that has been measured
by Kakiuchi and Senda at the water-nitrobenzene interface [136–138],
and the electrocapillary curves obtained are shown in Fig. 12. Similar
results were obtained by Girault and Schiffrin at the water/1,2-dichloro-
ethane interface [139].

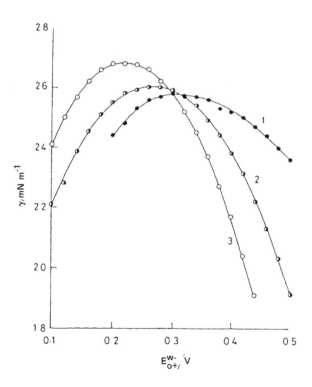

FIG. 12. Electrocapillary curves for the interface between nitrobenzene
solution of 0.1 mole dm^{-3} tetrabutylammonium tetraphenylborate and aque-
solution of (1) 0.01, (2) 0.10, and (3) 1.00 mole dm^{-3} lithium chloride at
25°C. The potentials refer to the cell: Ag|AgCl|0.1M TBACl(w)|0.1M
TBATPB(o)|σ|xM LiCl(w)|AgCl|Ag where σ represents the interface.
(Reprinted from Ref. 137 with permission. Copyright The Chemical
Society of Japan, Tokyo.)

It appears from Eq. [90] that if a physical plane of separation existed between the two phases, such as an ion-free layer of solvent molecules, Q would represent the net electrical charge density on the organic side of the plane, and this quantity would be equal and of opposite sign to the charge on the aqueous side. The assumption of a particular physical model can link the extensive variable Q obtained from the thermodynamic analysis to a spatially defined electrical charge distribution.

On the other hand, if such a plane of separation between the two phases could not be defined, as the water surface excess measurements tend to indicate, it becomes more difficult to link the measurable quantity Q to a specifically located physical charge. This is a consequence of possible interfacial mixing of solvent molecules and ionic species. Therefore, great care should be exercised in the analysis of quantities such as double-layer capacitance and its dependence on the free surface charges.

The absence of a physical plane of separation is also relevant to the study of biological systems such as cell membranes, where classical double-layer models have often been used without due regard to their applicability.

2. Differential Capacitances

From the Lippmann equation, it is possible to define the differential capacitance of the interphase by

$$C_d = \left(\frac{\partial Q}{\partial E} \right)_{\mu_{salt}, T, P} \tag{92}$$

C_d is a most readily accessible quantity in electrochemistry which can be measured by, for example, short-time galvanostatic steps, chronopotentiometry, or by ac impedance.

Extensive measurements of the differential capacitances of water-nitrobenzene and water/1,2-dichloroethane have been carried out by Samec et al. in order to test the Gouy-Chapman model adapted to the ITIES [140–147].

Girault and Schiffrin have shown that a double integration of measured interfacial capacitances from the potential of zero charge leads to
an electrocapillary curve identical to the one measured by interfacial
tension measurements [139].

3. *Esin-Markov Coefficients*

From the Gibbs equation other information can be obtained from the
cross-differential coefficients, such as, for example, the Esin-Markov
coefficient defined by

$$\left(\frac{\partial E}{\partial \mu_i}\right)_{\mu_{j \neq i}, Q, T, P} = -\left(\frac{\partial \Gamma_i}{\partial Q}\right) \tag{93}$$

Kakiuchi and Senda [137] have measured this coefficient for the water-
nitrobenzene interface and the results shown in Fig. 13 indicate the
apparent absence of specific adsorption for the aqueous salt LiCl and
organic salts TBATPB.

D.   Polarizable ITIES

The purpose of this section is to show the conditions for which an ITIES
can be polarized and to stress the differences between an ideally polarized
interface as described in Sec. IV.C and a real polarized one.

1. *Polarization Window*

The fundamental difference between the two cases is that an ideally
polarized interface refers only to an electrostatic equilibrium where the
constituents of each phase have an infinite standard Gibbs energy of
transfer, whereas a polarized interface always refers to a chemical equilibrium.   In the first case, when a potential difference is applied, no
current flows across the interface.   On the other hand, in the second
case, when a potential difference is applied, electrical current must flow
to establish the chemical equilibrium.   Consequently, if this current is
small enough to be considered negligible, the interface will be called
polarizable.   The definition of polarizability therefore implies the definition

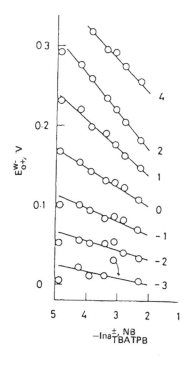

FIG. 13. Esin-Markov plots for nitrobenzene solution phase in contact with aqueous solution of 0.1 mole dm^{-3} lithium chloride. The surface charge density in aqueous phase is indicated by each line in $\mu C\ cm^{-1}$. Potentials as in Fig. 12. (Reprinted from Ref. 137 with permission. Copyright The Chemical Society of Japan.)

of a current threshold level under which a current is considered negligible, and the region for which this condition applies is called the polarization window.

In practice, and ITIES can be polarized if the aqueous electrolyte is very hydrophilic and the organic salt very hydrophobic. Let us consider as an example a system composed of an aqueous solution of 10^{-2} M LiCl and an organic solution of 10^{-3} M TPAsTPB in 1,2-dichloroethane, and let us calculate the size of the polarization window of this system.

At chemical equilibrium, the ionic concentration in both phases are related to the Galvani potential difference by Eq. (22):

$$\Delta_\alpha^\beta \phi = \Delta_\alpha^\beta {\phi}^{\,o} + \frac{RT}{z_i F} \ln \frac{a_i^{(\alpha)}}{a_i^{(\beta)}} \tag{94}$$

In the present case the ionic standard Galvani potentials are

$$\Delta_w^o \phi_{Li^+}^{\,o} \simeq -500 \text{ mV} \qquad\qquad \Delta_w^o \phi_{Cl^-}^{\,o} \simeq 481 \text{ mV}$$

$$\Delta_w^o \phi_{TPAs^+}^{\,o} \simeq 363 \text{ mV} \qquad\qquad \Delta_w^o \phi_{TPB^-}^{\,o} \simeq -363 \text{ mV}$$

If the Galvani potential difference $\Delta_w^o \phi$ is made very positive, we see that $TPAs^+$ will transfer from oil to water and Cl^- from water to oil.

On the other hand, if $\Delta_w^o \phi$ is made very negative, the aqueous concentration of $TPAs^+$ and the organic concentration of Cl^- can be made as small as to have only a satatistical significance.

Starting from very negative values of $\Delta_w^o \phi$, we see that when $\Delta_w^o \phi$ is made more positive, a current will flow in order to adjust the concentration of $TPAs^+$ in the aqueous phase and of Cl^- in the oil phase. The current associated with the transport of $TPAs^+$ from the organic phase, where the concentration of $TPAs^+$ is very small, let us say, 10^{-8} M, will be limited by the mass transfer in the water phase. The current is related to the mass transfer by

$$I_{TBA} = F k_m \, \Delta c_{TPAs^+} \tag{95}$$

where k_m is the mass transfer coefficient, usually of the order of 0.5×10^{-4} cm sec^{-1} for an unstirred system, and Δc_{TPAs^+}. the difference of concentrations of $TPAs^+$ between the interphase and the bulk.

If we assume for the present calculation that the interface is unpolarizable as long as the current necessary to establish the chemical equilibrium is smaller than 0.1 μA cm^{-2}, we see that for this threshold current the curface concentration of $TPAs^+$ will be 0.5×10^{-5} M. If we assume further that Eq. (94) can be used to measure the limit of the polarization window, we see that the current reaches the threshold level when the Galvani potential difference is equal to

$$\Delta_w^o \phi = \Delta \phi_i^o + \frac{RT}{z_i F} \ln \frac{a_i^{w \sigma}}{a_i^{o \sigma}} \approx 220 \text{ mV} \tag{96}$$

where a_i^σ are the activities of the ion i near the interface. Therefore, we can conclude that the interface is polarizable for potential differences more negative than 220 mV. A similar calculation shows that the interface is also polarized for $\Delta_w^o \phi$ values more positive than 220 mV, and that therefore the polarization window is about 440 mV for the example chosen. The results shown in Fig. 14 confirmed that the threshold current level chosen a priori and the estimated mass transfer coefficient were reasonable.

The reader should be warned that the discussion above has no rigorous thermodynamic foundation and assumes that during the transfer of ions, the local thermodynamic equilibrium and Eq. (94) prevail. Nevertheless, with these qualifications the method is very helpful in assessing the behavior of new systems.

2. Interfacial Ion Pair Formation

Another distinction between ideally polarized interface and real polarizable ITIES arises from the possible interfacial formation of ion pairs, as shown in Fig. 15.

Girault and Schiffrin [139] have shown that although an ion pair is a neutral species, ion pairs formed with an ion from each phase contribute to the electrical extensive variable Q, defined earlier as

$$Q = - \left(\frac{\partial \gamma}{\partial E} \right)_{\mu_{salt}, T, P} \tag{97}$$

Indeed, in the Gibbs equation, $\Gamma_i^{\alpha, \beta}$, the relative surface excess concentration of the species i, represents not only the free ionic species i but also the ionic species involved in an ion pair. It is therefore possible to write

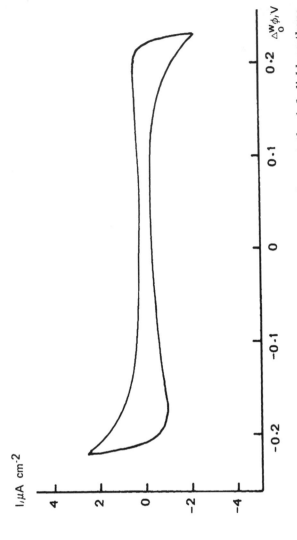

FIG. 14. Cyclic voltammetry for typical base electrolytes at the 1,2-dichloroethane (1,2-DCE)/water interface. Aqueous phase = 0.01M LiCl; organic phase = 10^{-3}M TPAsTPB; sweep rate = 20 mV s^{-1}.

$$\Gamma_i^{\alpha,\beta} = \Gamma_{i\,fc}^{\alpha,\beta} + \Gamma_{i\,ip}^{\alpha,\beta} \tag{98}$$

where $\Gamma_{i\,fc}^{\alpha,\beta}$ and $\Gamma_{i\,ip}^{\alpha,\beta}$ refer to the relative surface excess concentration of the free ions i and of the ion pairs containing the ion i, respectively.

Taking ion pair formation into account, the Gibbs equation obtained for the example of an ideally polarized interface [Eq. (89)] becomes

$$-d\gamma = \left(\Gamma_{K^+fc}^{\alpha,\beta} + \Gamma_{K^+ip}^{\alpha,\beta}\right) d\mu_{KCl} + \left(\Gamma_{A^-fc}^{\alpha,\beta} + \Gamma_{A^-ip}^{\alpha,\beta}\right) d\mu_{CA} + Q\,dE_{+(-)} \tag{99}$$

with

$$Q = F\left(\Gamma_{C^+fc}^{\alpha,\beta} + \Gamma_{C^+ip}^{\alpha,\beta} - \Gamma_{A^-fc}^{\alpha,\beta} - \Gamma_{A^-ip}^{\alpha,\beta}\right) \tag{100}$$

The extensive variable Q can then be written as the sum

$$Q = Q_{fc} + Q_{ip} \tag{101}$$

where Q_{fc} represents the free ionic charges and Q_{ip} represents the charge associated to the formation of interfacial ion pair.

OIL PHASE

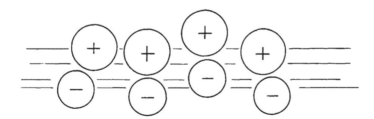

AQUEOUS PHASE

FIG. 15. Ion pairs present in the interfacial region make a contribution to the thermodynamic change Q (Eq. 98).

The free ionic charge Q_{fc} is identical to that considered in the case of ideally polarized interface and is given by

$$Q_{fc} = F \left(\Gamma^{\alpha,\beta}_{C^+ fc} - \Gamma^{\alpha,\beta}_{A^- fc} \right) = F \left(\Gamma^{\alpha,\beta}_{Cl^- fc} - \Gamma^{\alpha,\beta}_{K^+ fc} \right) \tag{102}$$

As mentioned in Sec. IV.B, interfacial ion pairs can be regarded as two specifically adsorbed ions, and the charge associated with the adsorption is Q_{ip}, equal in the present case to

$$Q_{ip} = F \left(\Gamma^{\alpha,\beta}_{KA\ ip} - \Gamma^{\alpha,\beta}_{CCl\ ip} \right) \tag{103}$$

where $\Gamma^{\alpha,\beta}_{KA\ ip}$ and $\Gamma^{\alpha,\beta}_{CCl\ ip}$ are the relative surface excesses concentration of the ion pair KA and CCl, respectively. Q_{ip} represents the number of ion pairs formed when the interfacial area is increased a unit amount. The interfacial formation of ion pairs at a liquid-liquid interface can be compared to the formation of hydrogen adatom on noble metals ($H^+ + e^- \rightleftharpoons H_{ads}$). The thermodynamics of these systems have been discussed extensively by Frumkin and co-workers [148], who also made a distinction between a free charge contribution and an ion pair contribution to the total charge.

If the interfacial ion pairs have a low solubility in either of the solvents, further complications in the investigation of polarizable ITIES arise from the interfacial precipitation of the salt. For example, with the system KCl in water-TBATPB in 1,2-dichloroethane, the salt KTPB is soluble neither in water nor in the organic solvent. Consequently, when a negative potential difference is applied, the precipitation of a white film of KTPB can be observed after a certain time [167].

E.   Potential of Zero Charge

The potential of zero charge is defined by

$$Q = \left(\frac{\partial \gamma}{\partial E_{+(-)}} \right)_{\mu_{salt}, T, P} = 0 \tag{104}$$

where Q is the extensive electrical variable of the interface. It has been shown that for an ideally polarized interface, Q is a thermodynamically well defined quantity, and in consequence, the Galvani potential difference at the potential of zero charge (pzc) has a unique value, although the actual value of the pzc measured depends on the reference electrodes chosen. In the absence of specific adsorption, the ionic contribution to the Galvani potential difference, $g_\alpha^\beta(ion)$, defined by Eq. (43) is equal to zero at this potential and therefore

$$\Delta_\alpha^\beta \phi_{pzc} = g_\alpha^\beta(dip) \tag{105}$$

The pzc is a readily accessible quantity that can be measured using a streaming electrolyte electrode, as illustrated in Fig. 16. The principle of the streaming electrode is that no charges can accumulate at the interface since its area increased continuously. Girault and Schiffrin showed that the pzc can be measured accurately and studied the system [149]

M	saturated calomel electrode	0.01 M KCl water	0.001 M TBATPB DCE	0.01 M TBACl water	saturated calomel electrode	M

O $\qquad\qquad\qquad\qquad\qquad\qquad\qquad\qquad\qquad$ E_{pzc}

The E_{pzc} measured for this cell was found to be equal to $-300\pm$ mV. In this case, the Galvani potential difference is related to E_{pzc} by

$$\Delta_w^o \phi_{pzc} = E_{pzc} + \Delta_w^o \phi_{TBA^+} \tag{106}$$

Equation (99) assumes that the diffusion potentials of the saturated calomel/aqueous electrolyte interface are negligible.

The Galvani potential difference between the TBATPB and the TBACl solutions, $\Delta_w^o \phi_{TBA^+}$, can be calculated:

$$\Delta_w^o \phi_{TBA^+} = \Delta_w^o \phi_{TBA^+} + \frac{RT}{F} \ln \frac{a_{TBA^+}^w}{a_{TBA^+}^o} \tag{107}$$

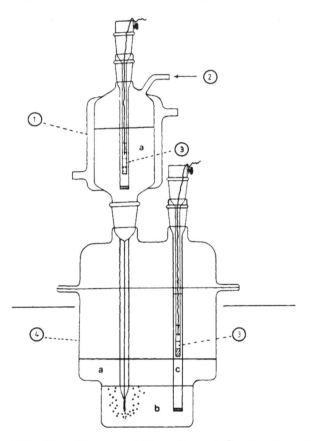

FIG. 16. Diagram of the cell used for the measurement of the point of
zero charge for an interface between two immiscible electrolyte solutions.
(a) Aqueous solution; (b) organic solution; (c) aqueous tetrabutylammonium bridge. (1) Jacketed aqueous solution reservoir; (2) nitrogen
pressure inlet; (3) saturated calomel reference electrodes separated from
the solutions by glass frits; (4) streaming electrode reservoir. (Reprinted
from Ref. 149 with permission. Copyright Elsevier Science Publishers
B.V., Amsterdam.)

Although the last two terms of Eq. (107) are not amenable to measurement, it is possible to estimate them provided that an extrathermodynamic assumption is made regarding the ionic standard Gibbs energies of transfer and that the limiting laws for the calculation of activity coefficients can be used. Using Czapkiewicz [91] ($\Delta G_{t,TBA^+}^{o,H_2O \to DCE}$ = -21.8 kJ mole^{-1}) and Abraham's data [93] (constant of association of TBA$^+$TPB$^-$ in 1,2-dichloroethane, K_A = 1.71×10^3), the Galvani potential difference at the potential of zero charge can be estimated to be 8 ± 10 mV.

The measurement of the pzc therefore gives an insight into the structure of the interface, as it gives an estimate of $g_w^o(dip)_{pzc}$. The very low value of $g_w^o(dip)_{pzc}$ obtained in the experiment described above shows that at the potential of zero charge the solvent-solvent short-range interactions are very small at the water-dichloroethane interface and consequently that there is no compact layer of oriented dipole molecules present at the interface. Similar conclusions have been reported by Samec [142] from capacitance measurements at the water-nitrobenzene interface.

F. Gouy-Chapman Model of the ITIES

The first extension of the Gouy-Chapman theory to an ideally polarized ITIES was proposed in 1939 by Verwey and Niessen [43]. Their model consisted simply of two back-to-back diffuse layers and was used by Ohlenbusch [46] in 1956 to interpret the results obtained by Kahlweit and Strehlow [44] at the water-quinoline interface. In 1977, Gavach et al. [129] proposed a refined version of this model assuming the presence of an ion-free layer composed of oriented solvent molecules separating two diffuse layers, as illustrated in Fig. 17.

Following the model considerations and the symbols used in Fig. 17, the Galvani potential difference between the two phases is given by

$$\Delta_\alpha^\beta \phi = (\phi_\infty^\beta - \phi_2^\beta) + (\phi_2^\beta - \phi_2^\alpha) + (\phi_2^\alpha - \phi_\infty^\alpha) \tag{108}$$

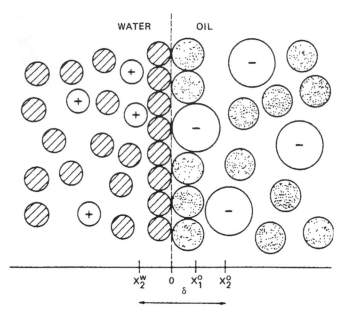

WATER   OIL

x_2^W 0 x_1^o x_2^o
      δ

FIG. 17. Schematic representation of the double layer at the water-nitrobenzene interface: $x_2^{W,o}$ are the coordinates of ions of the closest approach to the compact solvent layer (outer Helmholtz planes); x_1^o is the coordinate of an ion penetrating the inner layer over some distance $(x_2^W < x_1 < x_2^o)$. (Reprinted from Ref. 147 with permission. Copyright Elsevier Science Publishers B.V., Amsterdam.)

where $\phi_2 - \phi_\infty$ represent the potential drop in the diffuse layer, and $\Delta\phi_i = \phi_2{}^\beta - \phi_2{}^\alpha$ represents the potential drop across the inner layer.

The variation of $d\phi/dx$ in each diffuse layer is given by the classical Poisson-Boltzmann equation, which in the case of a 1:1 electrolyte reads

$$\left(\frac{d\phi}{dx}\right)_{x_2 < x < x_\infty} = \pm\sqrt{\frac{8RTc}{\varepsilon}}\, \sinh\left[\frac{F}{2RT}\,(\phi_x - \phi_\infty)\right] \tag{109}$$

Since the model assumes the existence of a physical barrier between the two phases, it is possible to define the charge of the diffuse layer by

$$\sigma = \pm\sqrt{8RTc\varepsilon}\, \sinh\left[\frac{F}{2RT}\,(\phi_2 - \phi_\infty)\right] \tag{110}$$

or

$$\sigma = \pm \, \varepsilon \left(\frac{d\phi}{dx}\right)_{x=x_2} \tag{111}$$

The electroneutrality of the interface $\sigma^\alpha = -\sigma^\beta$ leads to two interesting relations:

$$\varepsilon^\alpha \left(\frac{d\phi}{dx}\right)_{x=x_2^\alpha} = \varepsilon^\beta \left(\frac{d\phi}{dx}\right)_{x=x_2^\beta} \tag{112}$$

and

$$\frac{\sinh[F(\phi_2^\beta - \phi_\infty^\beta)/2RT]}{\sinh[F(\phi_2^\alpha - \phi_\infty^\alpha)/2RT]} = -\sqrt{\frac{\varepsilon^\alpha c^\alpha}{\varepsilon^\beta c^\beta}} \tag{113}$$

From Eq. (113), the potential drop in each diffuse layer can be calculated as a function of the difference $\Delta_\alpha^\beta \phi - \Delta\phi_i$, where $\Delta\phi_i$ represents the drop in the inner layer $(\phi_2^\beta - \phi_2^\alpha)$ by

$$\tanh \frac{F(\phi_2^\alpha - \phi_\infty^\alpha)}{2RT} = \frac{\sqrt{\varepsilon^\beta c^\beta} \, \sinh[F(\Delta_\alpha^\beta \phi - \Delta\phi_i)/2RT]}{\sqrt{\varepsilon^\beta c^\beta} \, \cosh[F(\Delta_\alpha^\beta \phi - \Delta\phi_i)/2RT] + \sqrt{\varepsilon^\alpha c^\alpha}} \tag{114}$$

It can be noticed that Eq. (112) expresses the continuity of the electrical displacement on both sides of the inner layer.

Using Eq. (114), it is therefore possible to illustrate the potential distribution at a liquid-liquid interface. Figure 18 shows the potential profile in the simplified case where $\Delta\phi_i = 0$ [170].

The usual way to test the Gouy-Chapman model is to compare the diffuse layer capacitance calculated from

$$C_2 = \frac{d\sigma}{d(\phi_2 - \phi_\infty)} = \pm \frac{2F\varepsilon C}{RT} \cosh\left[\frac{F}{2RT}(\phi_2 - \phi_\infty)\right] \tag{115}$$

with experimental values. The relationship between the measured values C and the values C_2 is given by

$$C^{-1} = \frac{d(\phi_\infty^\beta - \phi_\infty^\alpha)}{d\sigma} = \frac{d(\phi_\infty^\beta - \phi_2^\beta)}{d\sigma} + \frac{d(\phi_2^\beta - \phi_2^\alpha)}{d\sigma}$$

$$+ \frac{d(\phi_2^\alpha - \phi_\infty^\alpha)}{d\sigma} \tag{116}$$

or

$$C^{-1} = C_{2,\beta}^{-1} + C_i^{-1} + C_{2,\alpha}^{-1} \tag{117}$$

where C_i represents the inner-layer capacitance.

The fitting of experimental results to the Gouy-Chapman model at the water-nitrobenzene and water-dichloroethane interfaces was published in several communications by Z. Samac et al. [140–147], Buck et al. [150–151], and Kakiuchi and Senda [137].

Figure 19 shows a capacitance plot for a polarized water-nitrobenzene interface. It can be observed that the Gouy-Chapman values calculated for $\Delta\phi_i = 0$ represent a very good approximation at low concentrations. The present status of the study of the interfacial capacitance at the ITIES is still rather primitive. There are several problems that require

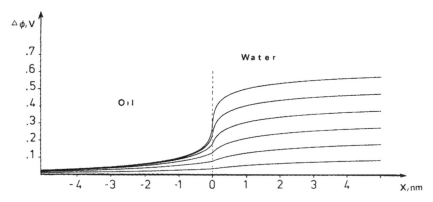

FIG. 18. Potential distribution across the interface between 10mM KCl in water and 1mM TBATPB in 1,2-DCE for different applied interfacial potentials. (Reprinted from Ref. 170 with permission. Copyright Elsevier Science Publishers B.V., Amsterdam.)

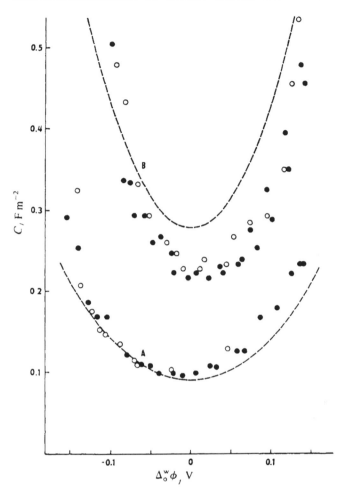

FIG. 19. Plot of the capacitance C of the water-nitrobenzene interface as a function of the potential difference $\Delta_o^W\phi$ for two concentrations of NaBr in water and of TBATPB in nitrobenzene: (A) 0.01 mol dm^{-3}, (B) 0.1 mol dm^{-3}. Capacitance data evaluated from the slope of the galvanostatic transient at t = 5 msec (filled circle) and t = 10 msec (open circle), Dashed lines show the capacitance of the diffuse double layer, C_d, calculated using Gouy-Chapman theory for $\Delta_o^W\phi_i$ = constant = 0 and relative permitivities ε^W = 78.54, ε^o = 34.82. (Reprinted from Ref. 144 with permission. Copyright The Royal Society of Chemistry, London.)

further work; most of the published work involves very dilute solutions
in the organic phase and in order to obtain information on possible inner
layers, large corrections due to the diffuse double-layer contributions
must be made. The theory currently used (Gouy-Chapman) is based on
a very primitive approximation which makes the separation of the measured
capacitance in individual components rather difficult.

Besides this problem, which can eventually be tackled using the
methods of molecular dynamics simulations, there is a more fundamental
difficulty in the modeling of the interface, in particular in relation to
the structure of the interfacial mixed-solvent region. The penetration
of this region by ions from the adjoining phases would appear to preclude
use of the simple ion-excluded inner-layer models proposed by Gavach
[129], Buck [215], and Samec [140].

VI.   CHARGE TRANSFERS
A.   Introduction

Different types of charge transfer across the interface between two
immiscible electrolyte solutions can occur via (1) ion transfer from one
phase to the other, or (2) electron transfer when a redox couple located
in one phase exchanges an electron with another redox couple located in
the other phase. These two phenomena have been studied experimentally
using standard electrochemical methodology, but the interpretation of the
results has not yet allowed a proper understanding of the kinetic mechanism. This is due to the fact that any theoretical approach of charge
transfer kinetics across an ITIES is primarily dependent on the physical
model of the interface used. As we discussed earlier (see Sec. V),
knowledge of the interfacial structure of ITIES has progressed since the
early model of Verwey-Niessen, and it is only recently that the concept
of a monotonic electrical potential profile suggested from the thermodynamic analysis has been generally accepted. The purpose of this
chapter is to present first the different theories proposed for charge

transfer and to illustrate the various electroanalytical applications. It should be pointed out that the mechanism of ion and electron transfer is still a matter of discussion, and therefore a more chronological approach will be followed to present the different ideas successively proposed.

B. Ion Transfer

1. Kinetics

After the pioneering work of Nernst and Riesenfield [1] in 1902 and the study of electroadsorption initiated in 1956 by Guastalla [47], the first investigation of ion transfer across a liquid-liquid interface using modern electrochemical methodology was carried out in 1972 by Gavach and Henry [152]. These authors studied by chronopotentiometry the transfer of tetrabutylammonium ion from an aqueous solution of NaBr to a solution of tetrabutylammonium tetraphenylborate in nitrobenzene. The general appearance of the chronopotentiograms obtained was similar to that generally obtained in electrode-electrolyte systems, and it was therefore concluded that the mechanism of ion transfer across a liquid-liquid interface was analogous to that for charge transfer on metallic electrodes. Consequently, the first theory of ion transfer proposed by Gavach [153] was a transposition of the theory usually employed to quantify electron transfer reaction on solid electrodes.

Assuming that the Galvani potential difference between the two phases $\phi(2) - \phi(1)$ is located entirely at the interface, Gavach et al. applied the Eyring-activated state theory to the ion transfer in a similar way to that commonly used in electrode kinetics. In this model, where the activated state ($\ddagger$) is considered to be in thermodynamic equilibrium with the initial (1) and final (2) stages, the phase transfer coefficient is defined by

$$\phi^{\ddagger} = \phi(1) + \alpha[\phi(2) - \phi(1)] \tag{118}$$

and the current-voltage relationship follows the Butler-Volmer equation, i.e.,

$$I = I_0 \left\{ \exp \left(\frac{-\alpha z F \eta}{RT} \right) - \exp \left[\frac{(1 - \alpha) z F \eta}{RT} \right] \right\} \tag{119}$$

where η represents the overpotential $\Delta_1^2 \phi - \Delta_1^2 \phi$ equilibrium and I_0 the exchange current density given by

$$I_0 = z F A k_0 [c_1]^{1-\alpha} [c_2]^\alpha \exp \left(\frac{-\alpha z F \, \Delta_1^2 \phi_i^o}{RT} \right) \tag{120}$$

where $\Delta_1^2 \phi_i^o$ is the standard Galvani potential of ion i defined by Eq. (21) and k_0 is given by

$$k_0 = \frac{kT}{h} \exp \left(\frac{-\Delta G_{act}}{RT} \right) \tag{121}$$

Further chronopotentiometric investigations [98,154−156] did not, however, confirm that ion transfer current-voltage characteristics followed a Butler-Volmer relationship rigorously. Gavach et al. [157] concluded that the proposed theoretical treatment should be regarded only as a first approximation.

In 1979, Gavach et al. [158] extended the theory above to fit the compact layer model of the interface illustrated in Fig. 17. In this model the Galvani potential difference between the two phases is distributed between two diffuse double layers and a central compact layer of oriented solvent molecules so that

$$\Delta_1^2 \phi = \phi_2(1) - \phi_2(2) + \Delta_1^2 \phi_c \tag{122}$$

where $\Delta_1^2 \phi_c$ represents the potential drop across the compact inner layer, and ϕ_2 the potential drop in each diffuse layer equal to the difference of Galvani potential between the plane of closest approach of ions to the interphase and the bulk of the solution. Using this model, Gavach [158] proposed to take into account a Frumkin correction for each phase, so that the rate constant is given by

$$k_{app} = k_0 \exp \left\{ -\frac{zF}{RT} [(1 - \alpha) \phi_2(1) + \alpha\phi_2(2)] \right\} \tag{123}$$

Experimental studies of the transfer of small tetraalkylammonium ions (C_2 to C_5) from water to nitrobenzene, together with the estimation of the diffuse layer potential drop ϕ_2 from a Gouy-Chapman analysis of interfacial tension data, showed that the changes of interfacial concentration with the applied potential were the cause of the potential dependence of the measured rate constants. On the other hand, discrepancies between the results and the model taking into account the Frumkin correction cast some doubts on the existence of the compact layer itself.

At about the same time, Samec [159] published a quantum mechanical basis to Gavach's theory following the Dogonadze treatment of electron transfer in chemical reactions. The treatment was based on the hypothesis that the activation energy for the transfer itself was given by

$$\Delta G_{act} = \frac{(\lambda + \Delta G_t)^2}{4\lambda} \tag{124}$$

where ΔG_t is Gibbs energy of the elementary transfer step, or in this case, the difference in the electrochemical potential of the ion between the two planes of closest approach:

$$\Delta G_t = \tilde{\mu}_i^2 - \tilde{\mu}_i^1 = zF(\Delta_1^2 \phi_i^o - \Delta_1^2 \phi) \tag{125}$$

where λ is the reorganization energy. This theory, which leads to the current-voltage relationship derived earlier by Gavach, did not, however, give any further mechanistic insight on the transfer process, as the author failed to expose the physical meaning of the solvent reorganization energy λ in this particular case of an ionic desolvation-resolvation process.

Using a four-electrode potentiostat [160], Samec et al. [161] investigated the transfer of Cs^+ ion by cyclic voltammetry. As in chronopoten-

tiometry, the shape of the cyclic voltamograms was similar to those
observed normally for reversible electron transfer reactions, but numerical analysis showed that the data did not fit the available theory exactly.
To explain the discrepancy between the ion transfer kinetic theory based
on the compact layer model and the experimental results, Samec et al.
[162] made use of convolution potential sweep voltammetry [163] to
analyze the potential dependence of the rate constant observed.

These measurements, coupled with the calculation of the potential
drop ϕ_2 in each diffuse double layer from a Gouy-Chapman analysis of
capacity measurements [140], showed clearly that the corrected rate
constant for the transfer of Cs^+ was independent of the applied Galvani
potential difference. Samec pointed out that this effect was due to the
fact that the potential drop across the inner compact layer between the
two planes of closest approach remained constant when the applied
potential was varied, as shown earlier by Gavach [99] and more recently
by Senda from electrocapillary measurements [137].

Buck and his group used a computer simulation technique to analyze
ion transfer kinetics data in relation to interfacial structure [164—167].
From the simulation of chronopotentiograms derived for the kinetic theory
of ion transfer proposed by Gavach for the different possible cases of
potential distribution, it was concluded that the ITIES was not suitable
for kinetic measurements because of the narrow potential window [114]
of these systems and the limited range of accessible ionic Gibbs energies
of transfer. In a more recent publication [168], they showed that the
measured ion transfer coefficient α is strongly dependent on current density
and nonaqueous supporting electrolyte concentration. One of the recent
important steps in the understanding of ion transfer is the work of
Senda et al. [169] in Japan. They studied the transfer of tetramethylammonium ion (TMA^+) from water to nitrobenzene by ac polarography.
Experimental results showed that not only the observed rate constant
but also the apparent charge transfer coefficient are potential dependent.

These observations therefore showed, as Gavach mentioned earlier, that the double-layer corrections alone could not explain the potential dependence of the ion transfer across a liquid-liquid interface.

In 1984, at the Heyrovsky Discussions held in Prague, most of the authors named above and the other participants agreed that the experimental evidence was in contradiction to the kinetics theory based on the compact layer model. The main question was: Is the elementary step of ion crossing across the compact layer a potential independent first-order chemical reaction, and can the current-voltage relation observed be explained only by the variation of the interfacial concentration at the plane of closest approach as the interface is polarized? In this case the current would be given by

$$ i = k_f c_i^{(1)} \exp\left[\frac{-zF\phi_2(1)}{RT}\right] - k_b c_i^{(2)} \exp\left[\frac{-zF\phi_2(2)}{RT}\right] \tag{126} $$

or

$$ i = k_f c_i^{(1)} \exp\left[\frac{-zF\phi_2^{(1)}}{RT}\right] \left\{ 1 - \exp\left[\frac{zF}{RT}\left(\Delta_1^2\phi - \Delta_1^2\phi_{eq}\right)\right] \right\} \tag{127} $$

which corresponds to a transfer coefficient equal to zero, which has never been observed experimentally. On the other hand, if the crossing of the compact layer is potential dependent, there is a contradiction with surface tension and capacity measurements which have clearly shown that the potential drop across the inner layer is very small and therefore cannot be considered a driving force for ion transfer.

In 1985, Girault and Schiffrin [170] proposed a new formalism for ion transfer kinetics at the ITIES. Starting from the fact that ion transfer is a transport process and not a chemical reaction, they applied Eyring-activated state theory for diffusion and migration [171] to the ion transfer process. From measurements of water surface excesses, they had proposed earlier [130] that the interface be considered as a mixed solvent region and therefore that the variation of standard ionic chemical

potential is monotonic within this interface. From electrocapillary data
[139], they had also proposed to regard the Galvani potential profile as
a continuous distribution between the two phases.

With this model of a mixed solvent region, the flux of an ion i in a
gradient chemical and electrical potentials between two equilibrium positions (a) and (b) located in the interfacial region near phases (1) and
(2), respectively, was given by

$$
j_i^{a \to b} = L \frac{kT}{h} c_i^{(a)} \exp\left(\frac{-\Delta G_a^{\ddagger}}{RT}\right) \exp\left(\frac{-\alpha_n}{RT} \Delta G_{t,i}^{o,a \to b}\right)
$$

$$
\times \exp\left(\frac{-\alpha_e zF}{RT} \Delta_a^b \phi\right) - L \frac{kT}{h} c_i^{(b)} \exp\left(\frac{-\Delta G_a^{\ddagger}}{RT}\right)
$$

$$
\times \exp\left[\frac{(1-\alpha_n)}{RT} \Delta G_{t,i}^{o,a \to b}\right] \exp\left[(1-\alpha_e)\frac{zF}{RT} \Delta_a^b \phi\right] \qquad (128)
$$

where $\Delta G_a^{\ddagger}$ is the activation energy of the "jump" from a to b, L the
distance between the two equilibrium positions, and α_n and α_e the
classically defined transfer coefficient relative to the chemical potential
and electrical potential gradients, respectively, such that

$$
\alpha_n = \frac{\mu_i^{o,\ddagger} - \mu_i^{o,a}}{\mu_i^{o,b} - \mu_i^{o,a}} \qquad (129)
$$

or

$$
\alpha_e = \frac{\phi^{\ddagger} - \phi^a}{\phi^b - \phi^a} \qquad (130)
$$

The concentration of the ion i at positions a and b, c_i^a and c_i^b can be
expressed as a function of the bulk concentration using a Boltzmann
distribution law:

$$
c_i^{(a)} = c_i^{(1)} \exp\left[\frac{-(\tilde{\mu}_i^a - \tilde{\mu}_i^1)}{RT}\right] \qquad (131)
$$

$$c_i^{(b)} = c_i^{(2)} \exp\left[\frac{-(\tilde{\mu}_i^b - \tilde{\mu}_i^2)}{RT}\right] \qquad (132)$$

The current is therefore given by

$$I = zFAk_0\left\{c_i^{(1)}\exp\left(-\bar{\alpha}_e \frac{zF}{RT}\Delta_1^2\phi - \bar{\alpha}_n \Delta G_{t,i}^{o,1\to2}\right) - c_i^{(2)}\right.$$

$$\left. \times \exp\left[(1 - \bar{\alpha}_e)\frac{zF}{RT}\Delta_1^2\phi + (1 - \bar{\alpha}_n)\frac{\Delta G_{t,i}^{o,1\to2}}{RT}\right]\right\} \qquad (133)$$

where $\bar{\alpha}_e$ and $\bar{\alpha}_n$, defined as the overall electrical and chemical transfer coefficients, are given by

$$\bar{\alpha}_e = \frac{\phi^{\ddagger} - \phi^1}{\phi^2 - \phi^2} \qquad (134)$$

$$\bar{\alpha}_n = \frac{\mu_i^{o,\ddagger} - \mu_i^{o,1}}{\mu_i^{o,2} - \mu_i^{o,1}} \qquad (135)$$

$\bar{\alpha}_e$ refers to the total Galvani potential difference between the two phases and not to the local interfacial potential drop. It represents the ratio of potential drop between the activated state and the bulk of phase (1) [nearly equal to the potential drop in the diffuse layer of phase (1)] to the total Galvani potential difference.

The overall transfer coefficient $\bar{\alpha}_n$, which is equal to the fraction of the Gibbs energy of transfer operative at the transition state, represents the degree of mixed solvation of the ion in that state. Assuming that this coefficient is potential independent, the current-voltage relationship for the ITIES is given by

$$I = I_0 \exp\left[\frac{-zF}{RT}(\bar{\alpha}_e - \bar{\alpha}_{e,eq})\Delta_1^2\phi_{eq}\right]\left\{\exp\left(\frac{-\bar{\alpha}_e zF\eta}{RT}\right)\right.$$

$$\left. - \exp\left[(1 - \bar{\alpha}_e)\frac{zF\eta}{RT}\right]\right\} \qquad (136)$$

where $\bar{\alpha}_{e,eq}$ represents the overall electric transfer coefficient $\bar{\alpha}_e$ at equilibrium and I_0 is the exchange current density, given by

$$I_0 = zFAk_0 \left[c_i^{(1)} \right]^{(1-\bar{\alpha}_{e,eq})} \left[c_i^{(2)} \right]^{\bar{\alpha}_{e,eg}}$$

$$\times \exp \left[\frac{-zF}{RT} (\bar{\alpha}_{e,eq} - \bar{\alpha}_n) \Delta_1^2 \phi_i^o \right] \tag{137}$$

It should be noticed that Eq. (137) differs from the classical Butler-Volmer relationship for electron transfer in two important respects: (1) the transfer coefficient is strongly concentration and potential dependent, and (2) the exchange current density depends on the value of the stand-ard transfer potential. Care therefore has to be taken in the direct transposition of electrode kinetic equations to the liquid-liquid interface case, and Eq. (137) should be employed.

2. *Electroanalytical Applications*

Although the kinetics of ion transfer has not yet been fully understood, experimental results have shown that the process of ionic transfer is fast enough to be classified as reversible. Therefore, electrochemical measurements can be performed under diffusion-controlled conditions to determine diffusion coefficients or to calculate thermodynamic quantities such as Gibbs energies of partition. The methodology associated with diffusion control ion transfer experiments is basically the same as that developed from Fick's diffusion law for metallic electrodes, and these techniques can be used at the ITIES since the boundary conditions are the same as those in classical electrochemical experiments. The principles of these electrochemical techniques can be found in any modern instru-mental electrochemistry textbook [172,173].

Until 1979, experimental investigations of the ITIES were restricted to controlled-current electrolysis since the instrumentation required in this case is limited to a constant-current source to apply the desired current between the two counter electrodes and a voltmeter to measure the interfacial potential difference between two reference electrodes. Such a simple setup has been used extensively by the group of Gavach at Montpellier [98,152—158] and later by the group of Buck in North Carolina [164,168] to carry out chronopotentiometric studies.

The second experimental approach used to investigate the ITIES was by impedance measurements between two large-area gold gauze electrodes located on both sides of the interface [155,174—177].

In 1977, Koryta et al. [178—180] studied the ITIES with an electro-lyte dropping electrode using a three-electrode configuration with the liquid-liquid interface located between a large-area working Pt electrode and a reference electrode for the aqueous phase. The considerable ohmic drop involved in this setup did not allow any quantitative analysis.

It was only in 1979 that Samec et al. [160,161] employed a four-electrode potentiostat and opened the way to all potential-controlled experiments. These four electrode potentiostats, which can be of dif-ferent design [181] and fitted with IR compensation, have been used successfully for cyclic voltammetry (Samec [161]), polarography (Samec [182]), differential pulse-stripping voltammetry (Marecek [211]), chrono-amperometry (Senda [183]), and ac impedance (Berlouis [185]). Another method recently introduced by Kihara et al. [186] for the study of the ITIES is current-scan polarography with analog subtraction of the IR drop from the observed interfacial potential difference.

An example of a typical cyclic voltammogram obtained at an ITIES is shown in Fig. 20 for the transfer of the picrate ions from water to nitro-benzene [190]. The usual criteria for reversibility apply to this system, i.e., independence of the peak potentials with sweep rate, ratio of anodic to cathodic peaks equal to 1 (after connection of the baseline of the reverse peak) [172], and a peak separation of 60 mV. Furthermore, the linear dependence shown in Fig. 21 of peak current on the square root of the sweep rate also indicates the diffusionally controlled character of this transfer and the applicability of the classical approach [172] to these systems.

Many other systems show identical behavior and, in fact, it appears that typical rate constants for ion transfer are too fast to be measured

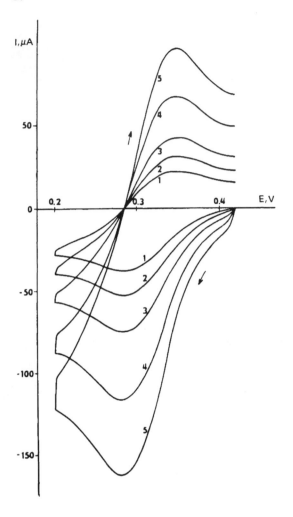

FIG. 20.   The voltammograms of the picrate ion transfer from the water to the nitrobenzene phase.   The picrate ion bulk concentration c_{Pi}^0 = 5×10^{-4} M, the interface area A = 1.77 cm^2, the polarization rates v: (1) 0.005 V sec^{-1}; (2) 0.01 V sec^{-1}; (3) 0.02 V sec^{-1}; (4) 0.05 V sec^{-1}; (5) 0.1 V sec^{-1}.   Potentials refer to the cell: Ag|AgCl|0.05M TBACl| 0.05M TBATPB + xM TBAP$_i$|σ|0.05M LiCl|AgCl|Ag; σ = interface; P$_i$ = picrate ion.   (Reprinted from Ref. 190 with permission.   Copyright Elsevier Science Publishers B.V., Amsterdam.)

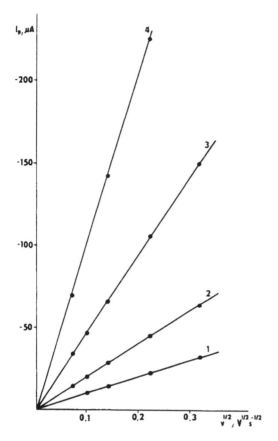

FIG. 21. The dependences of the peak current of the picrate ion transfer from the aqueous to the nitrobenzene phase on the square root of the polarization rate. Interface area $A = 1.77$ cm^2; the picrate ion bulk concentrations c_{Pi}^0: (1) 10^{-4}M; (2) 2×10^{-4} M; (3) 5×10^{-4} M; (4) 10^{-3} M. (Reprinted from Ref. 190 with permission. Copyright Elsevier Science Publishers B.V., Amsterdam.)

by this technique. The usefulness of cyclic voltammetry for studies at the ITIES is in the determination of half-wave potentials and diffusion coefficients.

Similar to the case of cyclic voltammetry, polarography at the ITIES has characteristics similar to those of the classical methods utilizing mercury. Again, the possibility of this transposition is due to the

identical boundary conditions, which are applicable in both cases. An
example of the results obtained by this technique is shown in Fig. 22,
where the transfer of tetramethylammonium ion at the water-nitrobenzene
interface was studied [182]. Again, a classical analysis of the polarographic wave based on the potential dependence of the current function log $[(i_l - i)/i]$ (i_l = limiting current) has been applied, and the
results corresponding to the data shown in Fig. 22 are given in Fig.
23. As can be seen, the classical theory [172] is applicable. From the

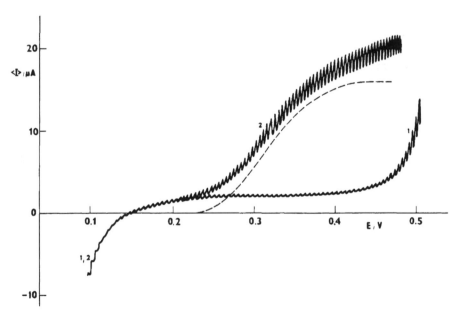

FIG. 22.   Original polarograms of the base electrolytes, 0.05 M LiCl +
1 M MgSO₄ in water and 0.05 M TBATPB in nitrobenzene, in the absence
(curve 1) and in the presence (curve 2) of 0.5 mM TMACl in the aqueous
phase.   Dashed curve, polarogram of TMA⁺ corrected for the base electrolyte current.   Rate of polarization 1 mV sec⁻¹.   The potentials cor-
respond to E = $\Delta_O^W \phi - \Delta_O^W \phi_{TBA^+}$ where $\Delta_O^W \phi_{TBA^+}$ is the Galvani potential
difference between 0.05M TBACl in water and 0.05M TBATPB in nitro-
benzene.   (Reprinted from Ref. 182 with permission.   Copyright Elsevier
Science Publishers B.V., Amsterdam.)

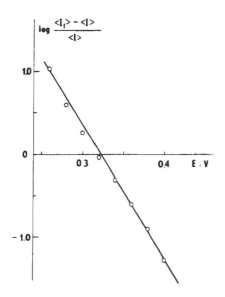

FIG. 23. Logarithmic analysis of polarogram of 0.5 mM TMA$^+$ corrected for the base electrolyte current. The slope of the straight line in 0.062 V; $E_{1/2}$ = 0.322 V. (Reprinted from Ref. 182 with permission. Copyright Elsevier Science Publishers B.V., Amsterdam.)

results shown in these figures, a value of the diffusion coefficient for TMA$^+$ of 6.6×10^{-6} cm^2 sec^{-1} was calculated, in agreement with previous determinations. The use of an ascending drop electrode presents some obvious advantages, as the surface is constantly renewed. Also, the drop time can be used as a measure of the interfacial tension, as it has been shown by Kakiuchi and Senda [137] that relative values can be obtained by this technique at the ITIES.

From the transposition of voltammetry to the study of the ITIES, it is possible to define an ionic transfer half-wave potential $\Delta_\alpha^\beta \phi_{i,1/2}$. This potential is related to the ionic standard Galvani potential $\Delta_\alpha^\beta \phi_i^o$ defined by Eq. (21) and is given by

$$\Delta_\alpha^\beta \phi_{i,1/2} = \Delta_\alpha^\beta \phi_i^o + \frac{RT}{zF} \ln \frac{\gamma_i^\alpha D_i^\beta}{\gamma_i^\beta D_i^\alpha} \tag{138}$$

where γ_i and D_i represent the activity coefficient and the diffusion coefficient of the ion. As shown before, the half-wave potential at the ITIES can be measured by the same methods as those used in classical electrochemistry, i.e., polarography and potential step chronoampero-metry, cyclic voltammetry (midpoint potential between the two peaks potential), or chronopotentiometry (potential at $t = \tau/4$).

One of the problems with studies at the ITIES is the high uncompen-sated ohmic resistance usually encountered in practice. For the accurate determnation of kinetic parameters, IR compensation techniques can be used conveniently. However, when compensation values greater than 1000 Ω have to be used for a potentiostatic measurement, the systems are often prone to oscillations. Furthermore, for potentiostatic experi-ments with ascending electrolyte drops, overcompensation can easily occur. To avoid these problems, Kihara et al. [186] used a current-scan technique in which the correction for the uncompensated cell resistance was accomplished directly by subtracting from the measured potential the IR drop across a resistance in series with the galvanostat. Some results obtained with this method are shown in Fig. 24, where the effect of a correct IR compensation can be clearly seen. Again, as observed previously by Samec et al. [182], a classical analysis is applic-able to this system (cf. Fig. 25).

An example of a chronoamperometric transient is shown in Fig. 26. It is difficult, however, with this technique to carry out measurements at short times due to the high values of the uncompensated cell resist-ance, and hence rate constants of ion transfer are difficult to measure. At constant time, the current-potential relationship is again, as expected for a reversible process, as shown in Fig. 27.

From the measurement of the half-wave potential it is therefore possible to estimate the ionic standard Galvani potential if assumptions are made regarding the activity and diffusion coefficients in each phase. The common assumption is to take the ratio of activity coefficients equal

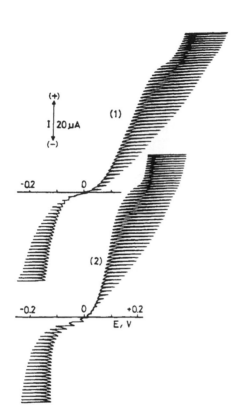

FIG. 24. Current-scan polarograms of TMA$^+$ at the aqueous electrolyte dropping electrode without (curve 1) and with (curve 2) IR drop compensation. Solution couple; 1×10^{-3} M TMABr, 0.05 M LiCl, and 1 M MgSO$_4$ in the aqueous EDE/0.05 M TBATPB in the nitrobenzene solution. (Reprinted from Ref. 186 with permission.)

to unity and to consider that the ratio of the diffusion coefficient is
equal to the reciprocal ratio of viscosity coefficients. Such an approach
proposed by Koryta [179], was followed by Osakai et al. [187] to determine the Gibbs energy of partition for 15 different alkylammonium ions,
from water to nitrobenzene. This method to determine Gibbs energies
of transfer relies entirely on the knowledge of ionic standard Galvani of
the ion used as a potential-determining ion for the reference electrode
of the organic phase. The cell used by Osakai et al. [187] was

$$
Ag \left| AgCl \left| \begin{array}{c} 0.1 \text{ M TBACl} \\ \text{water} \end{array} \right| \begin{array}{c} 0.1 \text{ M TBATPB} \\ \text{nitrobenzene} \end{array} \right\| \begin{array}{c} 0.1 \text{ M LiCl} \\ 5 \text{ mM RX} \\ \text{water} \end{array} \left| \begin{array}{c} 0.1 \text{ M LiCl} \\ \text{water} \end{array} \right| AgCl \left| Ag \right.
$$

The scale of Gibbs energy of transfer measured for the different ions
usually employed is relative to the Gibbs energy of transfer of TBA$^+$,
which was taken in this case equal to -0.248 V.

Despite the attraction of using a simple relationship between the
half-wave potential and the standard ionic transfer potential [Eq. (138)]
with some simplifying assumptions regarding the value of the activity

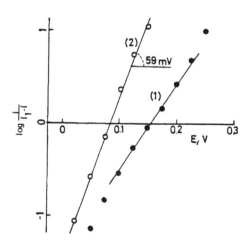

FIG. 25.   Test of the reversibility of TMA$^+$ transfer for the results in
Figure 24.

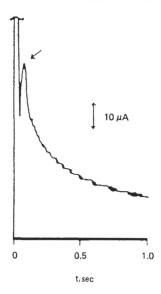

FIG. 26. Chronoamperometric current-time (i − t) curve of TMA⁺ ion transfer from the aqueous to the nitrobenzene phase at E = 0.38 V for 0.5 × 10⁻³ TMA⁺ in 0.1 mole dm⁻³ LiCl aqueous solution. A transient oscillation of the current (as indicated by an arrow) was sometimes observed, depending on the solution resistance and ion transfer impedance at the interface, at the initial part of the i − t record. Potentials with respect to the same cell as in Fig. 12; electrode area = 0.159 cm². (Reprinted from Ref. 183 with permission. Copyright The Chemical Society of Japan, Tokyo.)

coefficients, care must be taken with this analysis. This is particularly important for solvents of low dielectric constant, such as 1,2-dichloroethane. Problems arise from three sources: (1) the activity coefficient of large quaternary ammonium ions in aqueous solutions can differ considerably from unity in concentrated electrolytes, due to salting-out effects; (2) the activity coefficient of ions in nonaqueous solvents of low dielectric constant can differ significantly from unity even in dilute solutions (10^{-3} M) since the A constant in the Debye-Hückel theory is proportional to $\varepsilon^{3/2}$, and (3) ion pairing is common in the organic solvents usually employed, and this will lead to further lowering of the activity coefficients of the transferred ion.

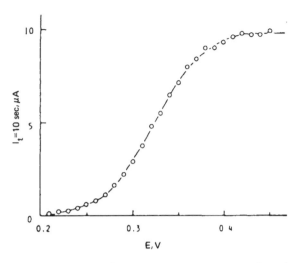

FIG. 27. Plot of I_t = 10 sec, corrected for the base current, against E for 1×10^{-3} mole dm^{-3} TMA$^+$ ion in 0.1 mole dm^{-3} LiCl aqueous solution. Potentials and electrode area as in Fig. 26. (Reprinted from Ref. 183 with permission. Copyright The Chemical Society of Japan, Tokyo.)

In view of these considerations, it is not surprising that some of the values of $\Delta\phi_i^o$ obtained from the measurement of the half-wave potential differ from the same quantity derived from data from extraction experiments. For example, from the data of Kakutani et al. [183], a value of $\Delta G_{t,TMA^+}^{o,w\to o}$ = 6.9 kJ mole^{-1} can be calculated. A similar result (6.2 kJ mole^{-1}) can be derived from the polarographic data of Samec et al. [182], and these data should be compared with a value of 3.4 to 4.0 kJ mole^{-1} obtained from partition experiment (see Table 1). The discrepancy is probably due to the neglect of the effects discussed previously. Despite these difficulties, these techniques are very powerful due to their simplicity.

To circumvent the problem of scales, Hundhammer and Solomon [188] proposed the use of tetraphenylarsonium tetraphenylborate as supporting electrolyte in the organic phase and to determine the zero of ionic standard Galvani potential in the TATB scale (see Sec. III.B) from cyclic and ac voltammetry. In this case the ionic standard potential obtained for different ions from the measurement of their half-wave potential is directly expressed in the TATB scale. Some results for the transfer of the ClO_4^- ion are shown in Fig. 28, where the results from the two techniques above mentioned are compared [184]. The ac current peak does not superimpose exactly for the forward and reverse sweeps as would be expected for reversible behavior, but no difference in the

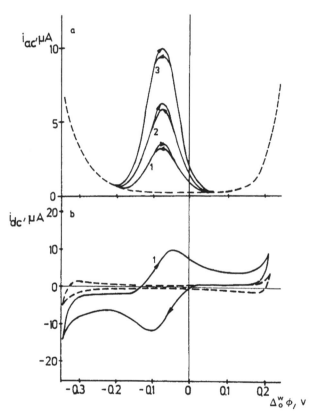

FIG. 28. Comparison of the dc cyclic voltammogram for the transfer of ClO_4^- across the water-nitrobenzene interface with the corresponding ac cyclic voltammogram (in-phase component output, f = 55 Hz) for three perchlorate concentrations: (1) 0.075 mM; (2) 0.15 mM; (3) 0.25 mM. Sweep rate 25 mV sec^{-1}. Dashed line denotes the supporting electrolyte. Aqueous phase = 0.01M Li_2SO_4; organic phase = 0.01M Crystal violet tetraphenylborate. (Reprinted from Ref. 184 with permission. Copyright Elsevier Science Publishers B.V., Amsterdam.)

peak potential can be seen. Also, a value of 90 mV for the half-peak
width is indicative of reversible transfer.

For electroanalytical applications, ac voltammetry is an interesting
technique, as it provides very low detection limits, of less than 10 μM,
and the possibility of simultaneous analysis of mixtures. An example of
this application is shown in Fig. 29, where it can be seen that the ions
NO_3^-, ClO_4^-, and Cs^+ can be analyzed together [184]. Ac impedance
analysis has also been used to study ion transfer, and the real and
imaginary components of the ac impedance for the transfer of the tetramethylammonium ion across the water-nitrobenzene interface obtained by
Osakai et al. [169] are shown in Fig. 30. The potentials refer to the
cell shown previously in this section. The same features of the ac voltammograms shown in Figs. 28 and 29 can, of course, be seen. The
interest of this work was, as mentioned before, the measurement of a

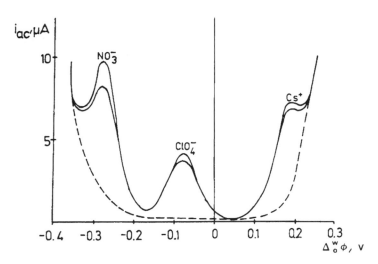

FIG. 29.   Ac cyclic voltammogram for the transfer of Cs^+, ClO_4^-, and
NO_3^- ions across the water-nitrobenzene interface.   Conditions as in
Fig. 28.   Concentration of NO_3^- and Cs^+ 0.2 mM and 0.1 mM ClO_4^-.
Dashed line denotes the supporting electrolyte.   Conditions as in Fig. 28.
(Reprinted from Ref. 184 with permission.   Copyright Elsevier Science
Publishers B.V., Amsterdam.)

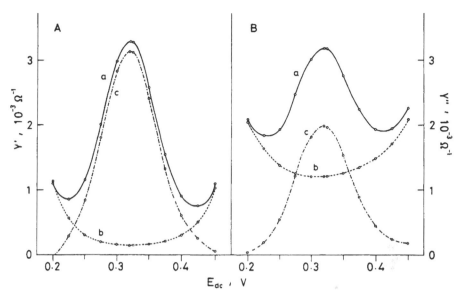

FIG. 30. (A) Real and (B) imaginary components of the admittance of the interface between a 0.1 mole dm^{-3} LiCl aqueous solution and a 0.1 mole dm^{-3} TBATPB nitrobenzene solution, at f = 50 Hz, in the presence (curve a) and the absence (curve b) of 0.5 × 10^{-3} mole dm^{-3} TMA$^+$ ion in the aqueous phase. Curve c shows the ion transfer admittance. Y_F^{ι} and $Y_F^{\shortparallel}$, for the TMA$^+$ ion transfer corrected for the base admittance. Potentials refer to the cell in Fig. 12. (Reprinted from Ref. 169 with permission. Copyright The Chemical Society of Japan, Tokyo.)

rate constant for the transfer of TMA$^+$, which is difficult to achieve by cyclic voltammetry.

Differential pulse stripping voltammetry has been used in the determination of acetylcholine by Marecek and Samec [211,214] and the results obtained are shown in Fig. 31. The technique involves a pre-electrolysis of 60 sec of acetylcholine into the organic phase followed by conventional differential pulse voltammetry. The resolution achieved is in the range 1 μM and the linearity in the concentration dependence of the stripping peak shown in Fig. 32 indicates the potential usefulness of these methods to complex systems.

The water-nitrobenzene interface has received much attention and many systems have been investigated. Apart from alkylammonium ions,

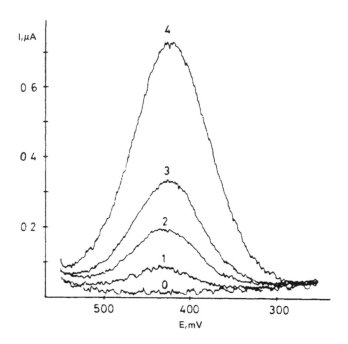

FIG. 31. Differential pulse stripping voltammogram of acetylcholine at
HEDE.   Concentrations of acetylcholine chloride in water (ppm): (0), 0;
(1), 0.5; (2), 1.0; (3), 2.0; (4), 5.0.   Aqueous phase = 0.01M LiCl +
xM acetylcholine; organic phase = 0.01M TPAsTPB.   The potentials are
measured with respect to a Ag AgCl 0.01M TPAsCl electrode.   (Reprinted
from Ref. 216 with permission.   Copyright Marcel Dekker Inc., New York.)

other ions have been studied, including choline and acetylcholine [101],
cesium [161,162,184], $Ru(bpy)_3{}^{2+}$, and MV^{2+} [189], and among the
anions: picrate [102,184,19], perchlorate [102,184,188], nitrate [184,188],
laurylsulfate [102], thiocyanate [102,188], tetrafluoraborate [188], and
iodide [191].

Besides the water-nitrobenzene interface, other interfaces have been
considered, especially the water/1,2-dichloroethane interface first proposed by Koczorowski and Geblewicz [192].   These authors studied by
chronopotentiometry the transfer of tetrabutylammonium ion [192], by
cyclic voltammetry the transfer of tetramethylammonium [193] and cesium
ion [194], and by ac impedance the transfer of picrate ion [195].   Ionic

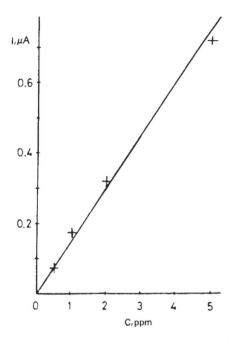

FIG. 32. Plot of the peak height of acetylcholine vs. concentration of acetylcholine chloride in water. Results from Fig. 31.

standard Galvani potentials for this interface have also been measured by Hundhammer and Solomon [188] by cyclic votammetry using tetraphenylarsonium tetraphenylborate as supporting electrolyte. It is interesting to notice that this organic salt gives a polarization window for this interface as wide as that obtained with water and nitrobenzene.

The water-isobutyl methyl ketone interface used in 1953 by Karpfen and Randles [30] to measure distribution potential was used by Koczorowski et al. [196] to study the transfer of tetraethylammonium, tetrapropylammonium, and iodide by both chronopotentiometry and cyclic voltammetry. Due to the narrowness of the potential window, Koczorowski showed that only these three ions could be studied, and consequently that this interface was not really suitable to invetigate ion transfer. Another liquid-liquid interface investigated is that between water and acetophenone. Solomon et al. [197] used dc and ac cyclic voltammetry to determine

standard Gibbs energy of transfer of a series of ions. These authors
showed that this interface does provide a reasonable polarization window
—not as large, though, as that obtained with water-nitrobenzene or
water/1,2-dichloroethane.

In addition to the four organic solvents used to form an ITIES,
preliminary results of the study of water-solvents mixtures have been
published for the following organic systems: nitrobenzene-benzonitrile
[198], nitrobenzene-benzene [198], and nitrobenzene-chlorobenzene [199].
Since selective solvation of ions in mixed solvent profoundly influences
many physiochemical phenomena, such as rates of chemical reactions,
solubilities, and so on, it is expected that the studies of solvent mixtures
could help the understanding of ion transfer or even electron transfer
mechanisms.

3. Assisted Ion Transfer

In the systems analyzed in Sec. VI.B.2, the only difference in the transfer properties of different ions was in their solvation properties. Thus
the standard transfer potential, which is the quantity that is all-important
in electroanalytical and separation science applications, was related only
to the Gibbs energy of solvation in both phases.

However, the presence of specific ion complexes in one of the two
phases is expected to result in significant shifts in the value of the
transfer potential, and the use of this idea in analytical applications has
received a considerable amount of interest in the last few years. We
consider here the thermodynamics of the case of a single uncharges complexing agent distributed between two phases, since this is the most
common situation studied:

$$
\begin{array}{ccccc}
M^{z+,\alpha} & + & mL^{\alpha} & \xrightarrow{\ K_a^{\alpha}\ } & ML_m^{z+,\alpha} \\[2em]
\Big\updownarrow K^{\alpha,\beta}_{\ M^{z+}} & & \Big\updownarrow K_1^{\alpha,\beta} & & \Big\updownarrow K^{\alpha,\beta}_{\ ML_m^{z+}} \\[2em]
M^{z+,\beta} & + & mL^{\beta} & \xrightleftharpoons[\ K_a^{\beta}\ ]{} & ML_m^{z+,\beta}
\end{array}
\qquad (I)
$$

where M^{z+} is the metal ion, L the specific ion ligand, m the stoichio-metric number of the complex formed, and $K_i^{\alpha,\beta}$ represents the distri-bution coefficient defined by Eq. (13). K_a is the homogeneous associa-tion constant equal to

$$K_a = \frac{a_{ML_m^{z+}}}{a_{M^{z+}} a_{L_m}^m} \qquad (139)$$

The important cross transfer reaction

$$M^{z+,\alpha} + mL^\beta \rightleftharpoons ML_m^{z+,\beta} \qquad (II)$$

is not included directly in the scheme of squares (I), as this and other cross reactions can always be expressed as a function of this general thermodynamic scheme. For the study of the kinetics of transfer, how-ever, the distinction between the different pathways is essential.

Since charge transfer reactions across the ITIES are in general reversible, the Galvani potential difference between the phases α and β is given by

$$\Delta_\alpha^\beta \phi = \Delta_\alpha^\beta \phi_{M^{z+}}^o + \frac{RT}{zF} \ln \frac{a_{M^{z+}}^\alpha}{a_{M^{z+}}^\beta} = \Delta_\alpha^\beta \phi_{ML_m^{z+}}^o + \frac{RT}{zF} \ln \frac{a_{ML_m^{z+}}^\alpha}{a_{ML_m^{z+}}^\beta} \qquad (140)$$

Koryta et al. [200–203] pioneered this new area of ion transfer, and in particular, its application to the study of the transfer of alkali metal ions from water to nitrobenzene in the presence of macrotetrolide anti-biotics such as nonactin and natural antibiotics such as valimomycin.

An example showing the facilitated transfer of Na^+ ion by the anti-biotic monensin is shown in Fig. 33. The coordination to this carboxylic ionophore shifts the transfer potential of Na^+ to within the polarization window of the ITIES at the nitrobenzene-water interface, whereas in the absence of this ionophore, the transfer of both Na^+ and TPB^- represents

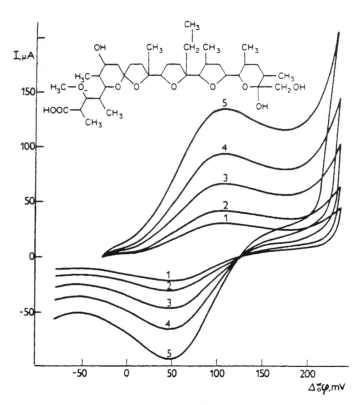

FIG. 33. Voltammograms of Na$^+$ transfer across the water-nitrobenzene interface facilitated by monensin. Aqueous phase, 0.1 M HCl, 0.1 M NaNO$_3$; nitrobenzene phase, 0.01 M TBATPB, 1 mM monensin. Polarization rates: (1) 5 mV sec^{-1}; (2) 10 mV sec^{-1}; (3) 25 mV sec^{-1}; (4) 50 mV sec^{-1}; (5) 100 mV sec^{-1}. (Reprinted from Ref. 204 with permission. Copyright Pergamon Press, Oxford.)

the negative limit of the polarization window for this system. Another example of the same behavior is shown in Fig. 34, but in this case, a synthetic crown ether is used as the specific acceptor for Na$^+$ ions [201]. In both cases, a diffusionally controlled transfer can be observed and in common with many other studies in this area, the concentration of the transferred ion in the aqueous phase is chosen much greater than that of the complexing agent.

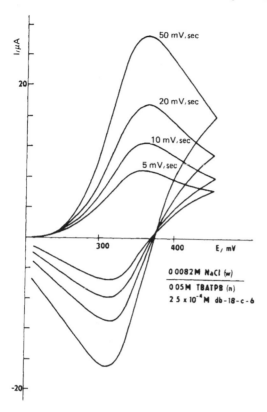

FIG. 34.   Voltammographic curves of the facilitated transport of sodium ion in the presence of 2.5 × 10⁻⁴ M db-18-c-6 at polarization rates of 5, 10, 20, and 50 mV sec⁻¹ in the 0.0082 M NaCl(w)/0.05 M TBATPB(n) system; T = 20°C. Potential scale the same as in Fig. 22. (Reprinted from Ref. 201 with permission. Copyright Elsevier Science Publishers B.V., Amsterdam.)

In these examples, we can consider the simplified equilibria

$$M^{+(w)} \rightleftharpoons M^{+(n)}$$

$$M^{+(n)} + L^{(n)} \rightleftharpoons ML^{+(n)} \tag{III}$$

for which the Galvani potential difference between the two phases is given by

$$\Delta_n^w \phi_{M^+} = \Delta_n^w \phi_{M^+}^o + \frac{RT}{F} \ln \frac{a_{M^+}^{(n)}}{a_{M^+}^{(w)}} \tag{141}$$

From Eq. 22,

$$\Delta_n^w \phi_{M^+} = \Delta_n^w \phi_{M^+}^o + \frac{RT}{F} \ln \frac{C_{LM^+}^n}{K_a C_L^n C_{M^+}^w} \tag{142}$$

where $\Delta_n^w \phi_{M^+}^{'o}$ is the formal potential for ion transfer, i.e., including activity coefficient terms. Since $C_{M^+}^{(w)}$ is in excess, only the diffusion of L and LM^+ in nitrobenzene need to be considered. At the half-wave potential the following condition applies:

$$\left(D_{LM^+}^{(n)} \right)^{1/2} \left(C_{LM^+}^n \right)_{x=0} = \left(D_L^{(n)} \right)^{1/2} \left(C_L^n \right)_{x=0} \tag{143}$$

and therefore the half-wave potential is given by

$$\Delta\phi_{1/2,LM^+} = \Delta_n^w \phi_{M^+}^{'o} + \frac{RT}{2F} \ln \frac{D_L}{D_{LM^+}} - \frac{RT}{F} \ln K_a C_{M^+}^w \tag{144}$$

From Eq. (144) it can be seen that the association constant K_a can be calculated from ion transfer experiments assuming that the diffusion coefficient of the ligand is the same as that of the complex. Furthermore, from cyclic voltammetry experiments it can be seen that this type of assisted ion transfer is completely diffusionally controlled, as the peak current depends linearly on the square root of the sweep rate and the ionophore concentration. Therefore, it can be used for the electroanalytical determination of ions (or conversely ionophores), which normally would transfer completely outside the potential window. An interesting application of this novel electroanalytical technique is in the determination of the alkaline earth cations using specific complexing agents. For instance, Marecek and Samec [97] employed cyclic ligands derived from 3,6-dioxaoctane/dicarboxylic acid to study the transfer of Ca^{2+} by differential pulse stripping voltammetry at a hanging electrolyte drop electrode. Figure 35 shows some results of this technique for different concentrations of Ca^{2+} using the selective ligand 7,19-dibenzyl-2,3-dimethyl-7,19-diazo-1,4,10,13,16-pentaoxacycloheneicosane-6,20-dione (PEDA). As can be seen, a resolution in the range 0.2 μM can be achieved (cf. Fig. 35).

The stoichiometry of assisted ion transfer is a matter of great practical importance in the area of industrial metal ions solvent extraction and the application of an electrochemical methodology to this problem is of great interest. For instance, Homolka et al. [205−207] showed the use of single-scan voltammetry for this purpose and derived a relation-

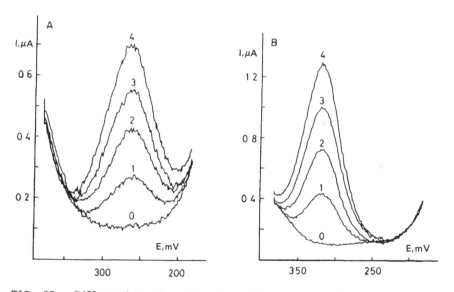

FIG. 35. Differential pulse stripping voltammograms of (A) calcium and (B) barium ion transfer at the HEDE. Concentrations (μM) of metal ion in the aqueous phase are indicated on the curves. Base electrolytes were 2.5 mM $MgCl_2$ in water and 5 mM TBATPB in nitrobenzene, with 1 mM ligand I in the nitrobenzene. The ohmic drop compensation was 6 kΩ. Potential scale similar to that in Fig. 22; experimental details as described in Ref. 161. (Reprinted from Ref. 97 with permission. Copyright Elsevier Science Publishers B.V., Amsterdam.)

ship between the peak shape and the stoichiometric number for the transfer of Li^+ and Cd^{2+} facilitated by the ligand PEDA.

The coupling of H^+ and ion transfer is a common occurrence in processes as diverse as biological membrane transfers and solvent extraction in hydrometallurgical applications. Koryta et al. [208,209] have studied the transfer of Na^+ ion in the presence of monensin, an acyclic complexant that forms cyclic complex with sodium and lithium by closing carboxylic and methoxylic group by hydrogen bonding. Monensin is a weak acid and its action in biological membranes is connected with the electroneutral exchange reaction of sodium and hydrogen ions given by reaction (III), with L = HX,

$$Na^{+(w)} + HX^{(m)} \rightleftarrows NaX^{(m)} + H^{+(w)}$$

HX = monensin and (m) refers to the membrane phase. At the nitrobenzene-water interface, monensin acts as an ion carrier, as can be seen from the cyclic voltammetry results in Fig. 33.

This is an interesting example of the usefulness of electroanalytical studies at the ITIES for the understanding of complex ion transfer processes. The half-wave potential for the monensin-Na^+-H^+ system was found to be

$$E_{1/2} = \frac{1}{2}(E_p^+ + E_p^-) = \frac{RT}{F} \ln \left(\frac{k_1}{C^w_{Na^+}} + \frac{k_2}{C^w_{H^+}} \right) \tag{145}$$

where E_p^+ and E_p^- are the peak potentials corresponding to the forward (water phase increasingly positive) and reverse scans, respectively, with k_1 and k_2 given by

$$k_1 = \frac{\exp(F \Delta_o^w \phi^o_{Na^+}/RT)}{K_1} \tag{146}$$

and

$$k_2 = \frac{\exp(F\Delta_o^w \phi_{H^+}^o/RT)}{K_2} \tag{147}$$

K_1 and K_2 are the stability constants for the formation of the monensin-Na^+-H^+ compounds given by

$$K_1 = \frac{a_{NHX^+}^{(n)}}{a_{Na^+}^{(n)} a_{HX}^{(n)}} \tag{148}$$

and

$$K_2 = \frac{a_{NHX^+}^{(n)}}{a_{H^+}^{(n)} a_{NaX}^{(n)}} \tag{149}$$

The pH dependence of the half-wave potential is whon in Fig. 36, from where it can be seen that the species NaX functions as the proton carrier at values of pH greater than ≈ 4.6, whereas the dissociated monensin, HX, is the sodium carrier at lower pH values. These con-clusions were confirmed by these authors from experiments at varying Na^+ ion concentrations. Equilibrium constant values for all the extrac-tion reactions could be calculated from the half-wave potentials and Eqs. (145) to (149). The values found were

$$Na^+ + HX \rightleftharpoons NaHX^+ \qquad \log K_1 = 5.63 \tag{IV}$$

$$H^+ + NX \rightleftharpoons NaHX^+ \qquad \log K_2 = 9.88 \tag{V}$$

and for reaction (III), $\log K = -4.54$. It is expected that this type of study will be of considerable interest in the elucidation of elementary steps in the ion transfer processes at biological membranes.

Facilitated proton transfers can be regarded as a special case of the specific ionophore transfers previously discussed. In this case, specific transfer is achieved through the protonation of a basic group in a

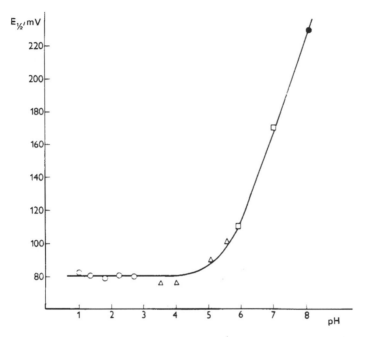

FIG. 36. Dependence of $E_{1/2} = E_p^+ + E_p^-)/2$ on pH of the aqueous electrolyte (at constant sodium concentration), with: dilute HNO_3 (open circle); acetate buffers (open triangle); phosphate buffers (open square); borate buffer (filled circle). Solid line, Eq. (145). Organic phase = 0.01M TBATPB + 1 mM monensin; aqueous phase = 0.1M Na^+ ion. (Reprinted from Ref. 209 with permission. Copyright Elsevier Science Publishers B.V., Amsterdam.)

hydrophobic molecule. This can be an amine or an N-heterocyclic compound [210] or a carboxylic function and it is to be expected that hydrogen ion transfer resulting from the acid-base properties of the acceptor will show high specificity compared with some inorganic cations. Considering the general case of a protonation reaction

$$A^{(n)} + H^{+(n)} \xrightleftharpoons{} AH^{+(n)} \tag{VI}$$

of the half-wave potential is given by

$$\Delta\phi_{1/2} = \Delta_w^n \phi_{H^+} + \frac{RT}{2F} \ln \frac{D_{AH^+}^{(n)}}{D_A^{(n)}} + \frac{RT}{F} \ln\left(K_{AH^+} a_{H^+}\right) \tag{150}$$

where K_{AH^+} is the protonation equilibrium constant [reaction (VI)].
Equation (150) corresponds to Eq. (144) for an acid-base reaction.
Thus, from the measurement of the half-wave potential, the values of
acid-base equilibrium constants can easily be calculated, provided that
the value of the free energy of transfer of the H^+ ions for the particular solvent is known.

Makrlik et al. [210] observed that facilitated protons transfer across
the nitrobenzene-water interface using 2,4-dinitro-N-picryl-1-naphthyl-
amine (α-hexyl) as the acceptor. Figure 37 shows the cyclic voltammetry
corresponding to this ion transfer. Similarly to the other systems discussed previously involving an ionophore, reversible behavior is observed, and a value of the acid dissociation constant of pK_a = 9.1 was
calculated, in reasonable agreement with a value of 9.75 obtained from
extraction data.

From Eq. (150) it can be seen that the range of values of acid
dissociation constants that can be measured using this method is limited
mainly by the potential window of the systems. It is interesting to
notice that a change of 1 pK unit in the acid dissociation constant will
result in a change of the half-wave potential of 60 mV. With the range
of hydrophobic base electrolytes currently available, it is expected that
pK values comprised between approximately 11 and 5 could be measured.
Homolka et al. [211] also studied the facilitated H^+ ion transport, using
aniline and o-phenanthroline as acceptors in the organic phase. Similar
voltammetric curves to those shown in Fig. 37 were observed for this
and for the transfer of other protonated amines dissolved in the aqueous
phase. Again, the values of the acid dissociation constants calculated
were similar to data obtained from extraction experiments, although some
consistent deviations were observed in all cases.

1,10-Phenanthroline and its derivatives were also used by Yoshida
and Freiser [212] to study proton transfers across the ITIES. These
authors used the ascending water drop electrode and a current-scan

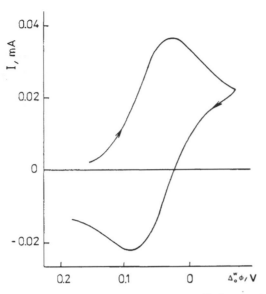

FIG. 37. Cyclic voltammogram of the proton transfer across the water-nitrobenzene interface (after a correction for the base electrolyte current) in the presence of 1.0 mM tetraphenylarsonium α-hexylate (TPAα) in the nitrobenzene phase. Base electrolytes: 0.01 M HCl in water, 0.01 M TPADCC in nitrobenzene. The polarization rate: 3.3 mV sec^{-1} (the arrows show the direction of the polarization). Interfacial area = 1.2 cm^2; E is the Galvani potential difference ($\phi^{(w)} - \phi^{(o)}$). (Reprinted from Ref. 210 with permission. Copyright Pergamon Press, Oxford.)

polarographic technique. Typical results of the pH dependence of assisted proton transfer can be seen in Fig. 38. These authors carried out a very extensive study to show that the electrochemical parameters measured were independent of the source of proton, either as the protonated compound or from an acid dissolved in the aqueous phase. The pH dependence for the half-wave potential for the various compounds studied is shown in Fig. 39, where the pH dependence predicted from Eq. (150) is seen to apply.

However, as the authors pointed out in this work, care has to be taken in the quantitative analysis of half-wave transfer potentials for solvents of low dielectric constant, since the nature of the transported

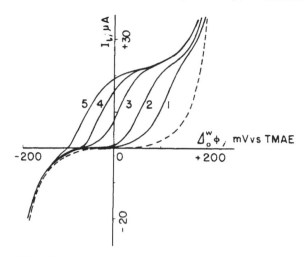

FIG. 38. Polarograms at an ascending water electrode in 1,2-dichloro-
ethane. Aqueous phase: 0.2 M acetate buffer. pH: (1) 6.7; (2) 6.1;
(3) 5.2; (4) 4.7; (5) 4.2. 1,2-Dichloroethane phase: 0.02 M TBA$^+$,
TPB$^-$ + 2.5 × 10^{-4} M 1,10-phenanthroline. The potentials are measured
with respect to a tetramethylammonium liquid membrane electrode. (Reprinted from Ref. 212 with permission. Copyright Elsevier Science
Publishers B.V., Amsterdam.)

species that is in equilibrium in the organic phase is not necessarily the
free ion. In fact, in 1,2-dichloroethane, extensive ion pairing is known
to occur [93] and the overall transfer reaction will be more closely
described by

$$A^{(n)} + H^{+(w)} + TPB^{-(n)} \rightleftharpoons AH^+TPB^{-(n)} \qquad (VII)$$

rather than by reaction (VI). In this case, the half-wave potential will
depend on the concentration of base electrolyte. For poorly dissociated
electrolytes such as TBA$^+$TPB$^-$ in 1,2-dichloroethane, the activity of
TPB$^-$ is given approximately by

$$c_{TPB} = \left[K_d c_{TBA^+TPB^-} \right]^{1/2} \qquad (151)$$

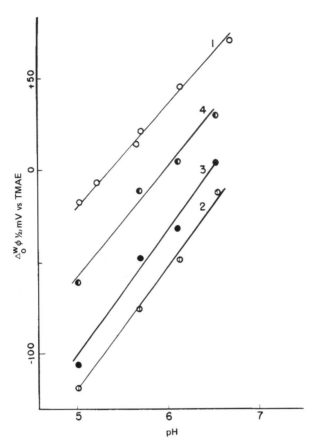

FIG. 39. Relation between $\Delta_o^a \phi_{1/2}$ and pH of aqueous phase. Aqueous phase: 0.2 M acetate buffer; 1,2-dichloroethane phase: 0.02 M TBA⁺TPB⁻ + 2.5 × 10⁻⁴ M. (1) 1,10-phenanthroline; (2) 4,7-dimethyl-1,10-phenanthroline; (3) 2,9-dimethyl-1,10-phenanthroline; (4) 4,7-diphenyl-1,10-phenanthroline. Results obtained in experiments similar to that shown in Fig. 28. (Reprinted from Ref. 212 with permission. Copyright Elsevier Science Publishers B.V., Amsterdam.)

where K_d is the dissociation constant of the electrolyte. When the species transferred form an ion pair with TPB^-, the half-wave potential relationship (150) will read

$$\Delta_w^o \phi_{1/2} = \Delta_w^o \phi_{H^+}^o + \frac{RT}{2F} \ln \frac{D_{AH^+TPB}^{(n)}}{D_A^{(n)}} + \frac{RT}{F} \ln (K_{AH^+} a_{H^+})$$

$$+ \frac{RT}{F} \ln (a_{TPB^-}^{(n)}) \tag{152}$$

From Eqs. (151) and (152) a TPB^- concentration dependence of ≈ 30 mV/decade should be expected for the half-wave potential, and this was indeed observed.

It can be concluded that although the electroanalytical techniques described are very powerful and convenient, the quantitative determination of equilibrium constants must take due account of the species transferred and, in particular, the possibility of ion pairing in the solvents usually employed in studies at the ITIES.

The transfer of protonated amines can be useful not only for the estimation of free energies of transfer, as discussed by Homolka et al. [211], but also as a tool for the quantitative determination of low concentrations of pharmaceuticals, such as tetracycline and its derivatives, as shown by Kozlov and Koryta [213]. Tetracycline is a zwitterion at neutral pH and a cation at pH values larger than the pK of the amino group, $-NH^+(CH_3)_2$ of the molecule ($pK_1 \approx 3.3$). Since the cationic form is only transferred, protonation prior to transfer will occur for pH values greater than 3.3. The half-wave potential is therefore given by [204]

$$\Delta_o^w \phi_{1/2} = \text{const.} - \frac{RT}{F} \ln \left(\frac{1}{K} + \frac{1}{a_{H^+}} \right) \tag{153}$$

The pH dependence observed was in agreement with Eq. (153). Similar analytical applications have been proposed by Homolka et al. [216] for

the case of the transfer of amines, and quantitative assays in the
micromolar range appear feasible.

The similarities of ion-selective electrodes and the electroanalytical
methods described must be stressed. Both use as an analytical criterion
the Gibbs energy of transfer as the discriminating factor between the
species that is being assayed and other species present. As can be
seen from the discussion in Secs. VI.B.2 and this one, the ion-selective
approach and electroanalysis appear to be complementary techniques,
and a great deal of information derived originally for ion-selective membranes has been transposed to electroanalytical applications of the ITIES.

C.   Electron Transfer

Electron transfer reactions have received a great deal of attention in the
last decade since these are the basis of many important chemical and
biochemical processes. The first studies focused on homogeneous reactions, and of particular importance was the study of self-exchange
reactions. In fact, most of the mechanistic concepts, such as inner-
sphere and outer-sphere reactions, are derived from this initial work
[217].

Marcus [218] extended the homogeneous adiabatic electron transfer
formalism to the heterogeneous case of electron exchange between a
redox system in solution and a metal electrode. Processes occurring in
nonhomogeneous environments are recognized as being of great practical
importance, especially those occurring in biological membranes. Liquid-
liquid interfaces have been proposed [219,220] as one of the simplest
approaches to the modeling of membrane chemistry since electron transfer reactions occur in living organisms between redox centers placed in
media of different polarity.

It is for this reason that studies of electron transfer at the ITIES
are of great interest, since these systems open up the possibility of
studying electron transfer reactions between redox centers with well-

defined relative positions and where the Galvani potential between redox centers is a controlled parameter. Therefore, these studies should link the two areas of homogeneous and heterogeneous electron transfers.

Electron transfer reactions at the ITIES had first been observed by Guainazzi et al. [221], who reduced aqueous Cu(II) to metallic copper using tetrabutylammonium hexacarbonylvanadate (−1) in 1,2-dichloroethane, by passing current through the interface. Subsequently, Samec et al. studied the electron transfer reaction between ferrocene in nitrobenzene and hexacyanoferrate(III) in the aqueous phase [222,162]. Unfortunately, no other studies of this important type of reaction have been reported in the literature.

1. Thermodynamics

We consider first the homogeneous electron transfer case:

$$\text{Ox } 1^\alpha + \text{Red } 2^\alpha \rightleftharpoons \text{Red } 1^\alpha + \text{Ox } 2^\alpha \qquad (VIII)$$

where Ox^α and Red^α mean the oxidized and reduced forms of each redox couple in the phase α. The degree of freedom of this reaction is equal to 1 because it involves four components related to each other by three restrictive conditions:

$$c_{O1}{}^\alpha + c_{R1}{}^\alpha = c_1{}^\alpha \qquad (154)$$

$$c_{O2}{}^\alpha + c_{R2}{}^\alpha = c_2{}^\alpha \qquad (155)$$

$$\Delta G^o_{O1/R2} = -RT \ln \frac{a_{O2}{}^\alpha a_{R1}{}^\alpha}{a_{O1}{}^\alpha a_{R2}{}^\alpha} \qquad (156)$$

In such an unvariant system, the choice of the concentration of only one of the four components determines the physical state of the system. At the liquid-liquid interface the situation is somewhat different, due to the presence of two distinct phases. The variance of the electron transfer interfacial reaction

$$O_1{}^\alpha + R_2{}^\beta \rightleftharpoons R_1{}^\alpha + O_2{}^\beta \qquad (IX)$$

between a redox couple O1/R1 located in phase α and a redox couple O2/R2 located in phase β is equal to 2, because in this case the number of variables is equal to 5, the concentrations of the four components and the Galvani potential difference, while the number of restricting conditions between them remains equal to the three conditions listed above.   In this case, the equilibrium condition is

$$\Delta G^{o}_{O1/R2} + RT \ln \frac{a_{O2}{}^{\beta} a_{R1}{}^{\alpha}}{a_{O1}{}^{\alpha} a_{R2}{}^{\beta}} + zF \Delta_{\beta}{}^{\alpha} \phi = 0 \tag{157}$$

where $\Delta_{\beta}{}^{\alpha}\phi$ represents the Galvani potential difference between the two phases.

For an ideally polarizable liquid-liquid interface it is possible to change the Galvani potential difference $\Delta_{\beta}{}^{\alpha}\phi$ by varying independently the concentrations of two components, since the variance is 2.   This is the main difference with the homogeneous case, where the concentrations of the components of the redox couples cannot be varied independently.

From Eq. (157) it is possible to define the standard Galvani potential of the electron transfer reaction as

$$\Delta_{\beta}{}^{\alpha}\phi^{o}_{O1/R2} = \frac{-\Delta G^{o}_{O1/R2}}{zF} = \frac{-1}{zF} [(\mu^{o}_{O2} + \mu^{o}_{R1}) - (\mu^{o}_{O1} + \mu^{o}_{R2})] \tag{158}$$

where μ^{o}_{i} is the standard chemical potential of species i.   Therefore, the Nernst equation for a redox reaction at an ITIES is

$$\Delta_{\beta}{}^{\alpha}\phi = \Delta_{\beta}{}^{\alpha}\phi^{o}_{O1/R2} + \frac{RT}{zF} \ln \frac{a_{O1}{}^{\alpha} a_{R2}{}^{\beta}}{a_{R1}{}^{\alpha} a_{O2}{}^{\beta}} \tag{159}$$

The standard Galvani potential for reaction (IX) is related to the standard redox potentials of the two redox couples expressed in a scale relative to the solvent α by

$$\Delta\phi^o_{O1/R2} = E^o_{O2/R2} - E^o_{O1/R1} + \frac{\Delta G^{o,\alpha\to\beta}_{t,R2} - \Delta G^{o,\alpha\to\beta}_{t,O2}}{zF} \tag{160}$$

where $\Delta G^{o,\alpha\to\beta}_{t,i}$ is the standard (ionic) Gibbs energy of transfer defined by Eq. (9). It can be seen that if an extra thermodynamic assumption regarding the Gibbs energies of transfer is made (see Sec. III.B), e.g., the ferrocene-ferricenium assumption proposed by Strehlow [79], the standard Galvani potentials can be referred back to the standard hydrogen electrode (SHE) in water. The other experimental approach is to measure the half-wave potential of the redox couple in the organic phase by a suitable voltammetric technique and refer the measured potential to the SHE by calculating junction potentials [162] using standard transfer quantities such as those shown in Table 1. Care should be taken in the latter approach in the calculation of the activity coefficients and possible ion pairing effects, due to the low value of the dielectric constant of the nonaqueous phases often employed.

2. *Kinetics of Electron Transfer*

Reaction (IX) is a special case of a bimolecular reaction, where the reactants and products are in different phases. The theory for this type of reaction was first discussed by Samec [159], who followed the usual approach in electrochemistry of considering an activated barrier to the act of electron transfer, the height of which is modified by the applied potential. Consideration of a particular interfacial model is inherent in this approach to account for the way in which the potential affects the different elementary reaction steps. For the bimolecular reaction (IX) Samec considered that two compact layers were present, one at each side of the interface. The Gibbs energy of activation can be calculated using the theory of Marcus for electron transfer reactions [218] from which [162]

$$\Delta G_{act} = \frac{(\lambda + \Delta G_{et})^2}{4\lambda} \tag{161}$$

where ΔG_{act} is the Gibbs energy of activation, ΔG_{et} the Gibbs energy charge associated with the act of electron transfer, and λ the solvent reorganization energy. Equation (161) is the starting point for the analysis of electron transfer reactions at liquid-liquid interfaces if these processes are regarded as an extension of the case of homogeneous electron transfers.

These ideas are not the only ones that could be used for the analysis of the kinetics of these reactions. For instance, the concept of redox levels in solution used by Gerischer [223] could be of use in this context.

The solvent reorganization energy has recently been calculated by Kharkats and Volkov [224] and, as expected, this quantity contains dielectric contributions from both media. These authors found that

$$\lambda = (ze)^2 \left[\left(\frac{1}{D_{op}^{\alpha}} - \frac{1}{D_{st}^{\alpha}} \right) \left(\frac{1}{2r_{O1}} - \frac{1}{2r} \right) + \left(\frac{1}{D_{op}^{\beta}} - \frac{1}{D_{st}^{\beta}} \right) \times \left(\frac{1}{2r_{R2}} - \frac{1}{2r} \right) \right] \tag{162}$$

where D_{op} and D_{st} are the optical and static dielectric constants, r_{O1} and r_{R2} the radii of the species O1 and O2, and r the sum of these ionic radii.

The potential dependence of the rate constant of electron transfer between ferrocene and hexacyanoferrate(III) at the water-nitrobenzene interface was studied by Samec et al. [162] using convolution potential sweep, voltammetry. If the liquid-liquid interface has an inner region, similar to, say, the mercury-solution interface, a strong potential dependence of the rate constant for electron transfer should be observed. However, the opposite appears to be the case [162]. The absence of a potential dependence in this case need not be conclusive proof of the interpretation of ions from both phases in the interfacial region, but may indicate, as suggested by Samec et al., the specific problems of highly charged ions in these studies (ion pairing, inconsistencies in Gouy-Chapman theory, etc.).

Girault and Schiffrin [225] have attempted to analyze this model of electron transfer using more recent ideas of the mechanism of the transfer process, employing the formalisms developed by Sutin [226]. The reaction is considered to occur through the sequence

$$O1 + R2 \underset{k_{-1}}{\overset{k_1}{\rightleftarrows}} \frac{O1}{O2} \underset{k_{-r}}{\overset{k_r}{\rightleftarrows}} \{O1R2\} \overset{\nu_{eff}}{\rightleftarrows} \{R1O2\} \underset{k'_r}{\overset{k'_{-r}}{\rightleftarrows}} \frac{R1}{O2} \underset{k_2}{\overset{k_{-2}}{\rightleftarrows}} R1 + O2$$

$$(X)$$

where O1/R2 and R1/O2 represent the precursor and the successor complexes, respectively; $\{O1R2\}$ the precursors complex reorganized to a configuration appropriate for electron transfer; and $\{R1O2\}$ the reorganized successor before relaxation to the state of successor complex. The effective hopping frequency is given by ν_{eff}.

The advantage of this approach is in the possibility of using a more detailed model of the interfacial region for the calculation of the equilibrium constants defining the formation of the precursor and successor, respectively, taking into account specific interactions such as interfacial ion pairing. This area of research has been little explored and basic information is still lacking.

VII. CONCLUDING REMARKS

The electrochemistry at the ITIES has advanced significantly in recent years and it is to be expected that this field will develop very rapidly, in particular in areas of application such as membranes for separation science, electroanalysis, solvent extraction, biomembrane studies, and so on. It is also of interest in the study of specific orientation effects of adsorbed amphiphilic molecules [220], since the possibility of applying potential differences across the interface results in specific orientation effects, as well as interfacial ionization. The amount of experimental

information is still comparatively small and further experimental work
is required.

APPENDIX: ELECTROCHEMISTRY OF LIQUID-LIQUID INTERFACES

Since this chapter was written in 1985, a great many new experimental
results and theoretical analyses on the electrochemical properties of immiscible electrolytes have been published. These are briefly discussed here.
The study of these systems has attracted a good deal of interest, particularly since the very wide range of applications to electroanalysis,
phase transfer catalysis, and solvent extraction has begun to be actively
investigated. Besides the obvious practical applications of these systems,
it has now been recognized that the study of charge transfer processes,
both ionic and electronic, across immiscible electrolytes gives an insight
into some more general problems of the physical chemistry of nonmetallic
electrified interfaces. Several reviews and books have been published
by Reid and co-workers [A1], Vaný́sek [A2], Kazarinov [A3], and
Girault [A4].

Equilibrium Interfacial Thermodynamics and Double-Layer Structure

Koczorowski [A5] has discussed the dipolar contributions to the Galvani
potential difference at the potential of zero charge according to the
analysis indicated by Eqs. (2.36–2.39). The data collected by Koczorowski indicated that $\Delta_o^W \phi_{pzc} \approx 0$ for the water/nitrobenzene interface.
Girault and Schiffrin [A6] proposed that, in the case of electrolyte systems that would not be expected to be specifically adsorbed at the interface, the value of the potential of zero charge could be used as an extra
thermodynamic assumption for establishing a scale of Gibbs energies of
ionic transfer if a value of zero was assigned to $\Delta_o^W \phi_{pzc}$. This assumption would still be valid in the case in which equal specific adsorption
of both the anion and cation of the organic base electrolyte occurred
through interfacial ion pair formation. The potential of zero charge has
been systematically studied by Koczorowski and co-workers [A7], using

the streaming electrode method previously described [149] for 1,2-DCE, nitroethane, nitrobenzene, and nitrobenzene/benzene mixtures containing TBATPB in contact with aqueous LiCl. As previously observed, the value of $\Delta_o^w \phi_{pzc}$ was in nearly all cases more negative (~40-80mV) than the potential of the minimum of the double-layer capacitance measured by either ac or galvanostatic techniques. The understanding of the origin of this difference has an important bearing on the theory of the double layer of these interfaces. Koczorowski et al. [A7] discussed the possibility that this difference was related mainly to the dynamic character of a measurement with a streaming electrode, both from the point of view of slow adsorption equilibrium or of nonideal polarizability. These questions still require elucidation, and it is quite clear that the equivalent circuit of an immiscible electrolyte interface is more complex than a simple series capacitance and resistance combination, as shown by Samec et al. [A8]. These authors studied the double-layer capacitance of the 1,2-DCE/aqueous electrolyte interface by subtracting the Warburg impedance contribution to the measured impedance and they concluded that, although no surface excess measurements were carried out, no evidence for specific adsorption of ions or ion pairs could be found. This work shows very clearly, however, the complete breakdown of a classical double-diffuse double-layer model coupled to ion free layers, i.e., to an "inner"-layer interfacial capacitance. If this model is adopted, negative inner-layer capacitances are obtained for the dilute solutions. This may be a consequence of the inadequacy of the classical Gouy-Chapman (G.C.) theory, as the authors recognized and has been clearly shown to be the case by Torrie and Valleau [A9]. These latter authors carried out a Monte Carlo simulation of two diffuse double layers, taking the image potential into account, and found that the classical G.C. theory can give very misleading results. In view of these uncertainties, it is difficult at present to ascertain whether an "inner" layer, along the lines of the classical Hg/solution interface, is realistic for these interfaces. Certainly, the evidence that can be derived from the potential dependence of ion transfer [A10,A11] indicates that such a model is at variance with experimental observations.

Specific adsorption at immiscible electrolyte interfaces has been observed for large hydrophobic ions such as hexdecyltrimethylammonium, $HTMA^+$ (Kakiuchi, Kobayashi, and Senda [A12]), or hydrophobic derivatives of affinity dyes, such as cibacron blue and procion blue (Schiffrin et al. [A13]). Actually, it is important to notice that, in both cases, the ion that is specifically adsorbed has a polar head and a hydrophobic residue. In the case of $HTMA^+$, the $-N^+(CH_3)_3$ residue has a strong interaction with the aqueous phase owing to its small size, and it is not surprising, therefore, that strong specific adsorption was observed when the organic phase was positively charges [A12]. A clear example of this phenomenon is the adsorption of amphiphiles [220], and Wandlowski et al. [A14] presented some preliminary capacitance measurements in the presence of adsorption of 1,2-dipalmitoyl-sn-glycero-3-phosphatidylcholine at the nitrobenzene/aqueous solution interface. The adsorption of phosphatidylcholine and phosphatidylethanolamine at the water/1,2-DCE interface shows a clearly defined potential dependence indicative of adsorption-desorption phenomena caused by the alteration of the acid-base equilibrium at the interface by the applied potential [220].

Girault and Schiffrin [A15] showed that there is a strong coupling among the pH, the interfacial potential, and the interfacial tension of phospholipid monolayers and that this coupling gives rise to surface tension motility phenomena. It has been further proposed that these effects culd be useful for the understanding of elementary motility processes in living organisms. Further work is required to study these elementary cell recognition processes. Movements induced by changes in the interfacial tension at liquid-liquid interfaces under applied alternating potentials have also been observed by Makino and Aogaki [A16].

The adsorption of the TPB^- anion at the water-air and at the nitrobenzene-saturated water-air interfaces has been studied by Zagórska

and Koczorowski [A17]. TPB⁻ is strongly adsorbed, as would be
expected for a large organic ion, but the simultaneous presence of
nitrobenzene in the adsorbed layer results in a decrease of the magnitude of $\Delta\chi$, indicating an antiparallel orientation of the adsorbed dipoles
with respect to the adsorbed TPB⁻Na⁺ pair.

Gibbs energies of transfer are easily measured at liquid-liquid interfaces by standard electrochemical methods. Until now, the most successful approach for rationalizing these quantities has been that due to
Abraham and Liszi [86—90], who proposed a simple two-layer dielectric
model, with the thickness of the first solvation layer determined by the
molar volume of the solvent. In this model, the dielectric constant of
this layer is taken as 2. Kornyshev and Volkov [A18] and Markin and
Volkov [A19] calculated the effects of dielectric saturations close to an
ion by choosing a Langevin function to describe these polarization effects
on the local permittivity. As expected, these authors found that dielectric saturation makes a significant contribution to the ionic solvation
energy. The difference between these results and those of Abraham is
not clear; the successful model of Abraham uses a cutoff approximation
instead of a dielectric function, the physical meaning of which can be
regarded as due to the saturation effects discussed in this paper.
Both models really treat the liquid as a continuous dielectric, and it
would seem that advances in the understanding of the nature of the
solute-solvent interactions require not the refinement of continuum
models but rather the use of theories of the liquid state.

Ion Transfer

The study of the kinetics of ion transfers across immiscible interfaces
has attracted a great deal of attention for several reasons. First, the
theory of ion transfers across immiscible solutions is a natural extension
to the problem of transport in homogeneous media, but with the added
complication of unidimensional inhomogeneity. Second, solvent extraction
in hydrometallurgy is in fact a particular case of simultaneous ionic

transfers and, finally, there are many practical analytical applications
of ion transfer using assay techniques that are identical to well-established electroanalytical procedures.   Cunningham and Freiser [A20] have
computerized the measurements of current polarography at liquid-liquid
interfaces and have incorporated in the same instrument accurate drop-
time measurements for calculating the interfacial tension.

The transfer of TMA^+, TEA^+, TBA^+, Cs^+, and picrate ions across
the 1,2-DCE/water interface was qualitatively studied by Geblewicz and
co-workers [A21] by cyclic voltammetry and found to be reversible.
The distribution potential of TEAPi is close to zero and, hence, as these
authors previously proposed, the use of this solute could provide a
practical way of making a zero Galvani potential junction between the
organic and the aqueous solution for measurements of standard transfer
potentials.   The transfer of TBA^+ has also been found to be reversible
by Pang et al. [A22].

Most of the work published to date has been using 1,2-DCE and
nitrobenzene as organic solvents because of their low miscibility with water
and high dielectric constants.   It is useful though to have a wider range
of solvents in order to ascertain the generality of the results obtained.
Alemu and Solomon [A23] studied ionic transfers across the o-nitro-
toluene/water interface, and the results obtained are given in Table A1.
The measured standard Gibbs energies of transfer were analyzed using
Abraham's solvation model [86–90]. $\Delta G_{t,i}^o$ for ClO_4^-, IO_4^-, and TBA^+ was
closer to theoretical predictions assuming no hydration in the organic
phase, whereas the value of $\Delta G_{t,TEA^+}^o$ indicated hydration.   These considerations appear to favor the retention of water of hydration in the
organic solvent on transfer from the aqueous phase and have important
consequences for the understanding of the kinetics of ion transfer.
Independent spectroscopic evidence is necessary in order to understand
the state of ions in water-saturated organic solvents that are immiscible
with water.   The relationship between the nature of the solvent and the
available potential window has been extensively studied by Kihara et al.

TABLE A1

Values of $\Delta_O^W \phi^o$ for Different Ions and Solvents Immiscible with Water

Ion/Solvent	O-Nitrotoluene[a]	1,2-DCE[b]	1,2-DCE[c]	Nitrobenzene[d]	Chloroform[c]	Nitrobenzene[d]
TBA+	-0.197	-0.206				
TEA+	-0.005	0.042				
TMA+	0.087	0.198				
TPAs+	-0.256					
ClO4-	-0.069		-0.17	-0.05	-0.19	
SCN-	-0.115					
IO4-	-0.066		-0.17	-0.06	-0.19	-0.087
TPB-	0.267					
Cs+		0.185				
Pi-[e]		-0.030				
ReO4-			-0.15			-0.077
MnO4-						-0.016
BF4-			-0.19	-0.10	-0.24	
I-			-0.25	-0.19	-0.21	
Br-			-0.37	-0.31	-0.33	

TABLE A1 (Continued)

Ion/Solvent	O-Nitrotoluene[a]	1,2-DCE[b]	1,2-DCE[c]	Nitrobenzene[d]	Chloroform[c]	Nitrobenzene[d]
ClO_3^-			-0.33	-0.27	-0.31	
BrO_3^-			-0.39	-0.34	-0.37	
NO_3^-			-0.33	-0.27	-0.29	
ClO_2^-			-0.41	<-0.35	-0.40	
NO_2^-			-0.41	<-0.35	-0.37	
$Cr_2O_7^=$			-0.32	<-0.35	-0.31	
IO_3^-			<-0.45	<-0.35		
$MoO_4^=$			<-0.45	<-0.35		
$WO_4^=$			<-0.45	<-0.35		
Cl^-			<-0.45	<-0.33	-0.41	

[a]Ref. A22; [b]Ref. A21; [c]Ref. A25; [d]Ref. A27; [e]picrate.

[24] using an electrolyte dropping (or ascending) electrode and taking the limits of the polarization window as the potential corresponding to a current of 20 μA. The experimental conditions used by these authors (flow rate = 0.013 ml sec^{-1}, drop time = 6.3 sec) corresponded to current limits of ±22 μA cm^{-2}. These currents are quite high for some of the solvents used to ensure that the results are independent of ohmic drops. It is interesting though that there is a very wide range of solvents that can be used for electrochemical studies at liquid-liquid interfaces. Some of the results obtained are given in Table A2. When the potential window is determined by the transfer of the organic electrolyte, its value can be greatly altered by salting out, using concentrated solutions of either $MgSO_4$ or Li_2SO_4, as shown by Geblewicz et al. [A25]. In fact, these authors could correlate the change of limits of polarization to the size of the organic ions using salting-out theories, and polarization windows in excess of 0.7 V could be obtained.

TABLE A2

Potential Window for Potential Solvents for Two Different Organic Salts. Aqueous Solution = 1 M $MgSO_4$

Solvent	0.05 M CV^+TPB^-[a]	0.05 M $TPAs^+DA^-$[b]
Chloroform	0.71	0.59
Aniline	0.24	—
o-dichlorobenzene	0.68	0.58
1,2-dichloroethane	0.74	0.66
o-chloroaniline	0.54	0.50
m-chloroaniline	0.51	0.46
Bis-(2-chloroethyl) ether	0.65	0.68
1-nitropropane	0.64	0.52
Benzonitrile	0.60	0.49
2-nitropropane	0.62	0.60
Nitrobenzene	0.66	0.62

[a]Crystal violet tetraphenyborate; [b]tetraphenylarsonium dipicrylaminate.

The potentials if ion transfer from water to different immiscible solvents has been measured by Kihara and co-workers [A24,A26,A27] and a summary of their data, as well as that obtained by other authors, is given in Table A1. The values of $\Delta G^o_{t,i}$ were correlated to a simplified Born type of equation, but the agreement was rather poor, reflecting the need to consider in more detail the nature of the first solvation layer round the ions.

Osakai and co-workers [A10] studied the kinetics of transfer of the picrate ion across the nitrobenzene/water interface by ac polarography. The rate constants for ionic transfer were measured for different concentrations of electrolytes. When these values were corrected according to the Frumkin formalism based on the assumption that the reaction planes are located on the outer Helmholtz planes (OHP) of the respective solutions, the corrected rate constants were found to be strongly dependent on the base electrolytes concentrations. This inconsistency does not result from ion pairing or activity coefficient effects and, therefore, the classic model of inner and diffuse layers for the interfacial structure is not valid for these interfaces. A similar problem with the analysis of the rate constant for the transfer of tetraalkylammonium ions was noticed by Samec and Mareček [A11]: after correcting for diffuse double-layer potentials, the rate constant was found to be potential independent, i.e., the "inner" potential difference (if present) is independent of the applied potential. These results reinforce the view that a mixed solvent layer is a more appropriate model for the interfacial region. It is interesting that a Bronsted-type relationship between the potential independent rate constants and the Gibbs energies of transfer was found, indicating that the use of the concept of a "chemical" transfer coefficient [A33] is correct.

The measurement of rate constants for ionic transfer in these systems is complicated by the high ohmic resistances usually encountered, which can easily result in large errors in the measured rate constant. In order to avoid these problems, Taylor and Girault [A29] developed a microelectrode technique for use in immiscible electrolyte systems. It was shown that this technique, similarly to the behavior of metallic microelectrodes, offers great advantages for the measurement of fast transfer kinetics. The difference with metallic electrodes is that a combined linear-spherical diffusion situation is present that is reflected in the unusual shape of the cyclic voltammograms. This does not introduce any serious problem for extracting analytical and kinetic data, as shown by computer simulation studies [A30].

The application of studies on the kinetics of individual ion transfer to the important practical case of salt extraction has been described by Koryta [A31] and Koryta and Skalický [A32]. The basis for the approach is similar to the basis for the theory of corrosion processes: when a salt is extracted from one medium into another, the flux of the cation must equal that of the anion; otherwise, electroneutrality would be violated. An extraction potential can be calculated in a way similar to that used for the prediction of a corrosion potential from the kinetics of the individual anodic and cathodic reactions, provided the rate constants and the form of the kinetic equation are known. This latter problem has been analyzed by Girault and Schiffrin [A33] using an extension of the theories of homogeneous ion transport. The rationale behind this approach is the observation that the Gibbs energies of transfer of many ions are comparable to the activation energy of homogeneous ionic transport. Hence, if the interfacial region can be regarded as a mixed solvent layer, it is reasonable to expect that the formalisms describing homogeneous ion transport can be extended to the liquid-liquid interface case. A different approach has been taken by Gurevich and Kharkats [A34]. In common with other approaches [A33], it is considered that the rate of transfer will be determined by a potential energy barrier that can be altered by the applied interfacial potential. Gurevich and Kharkats assumed that the potential energy curve near the energy maximum describing the transfer on the ion was parabolic with respect to the reaction coordinate. The application of this model to the calculation of rate constants of ion transfer has been presented [A35] and, from this elaborate treatment, it was concluded that the existence of a potential energy barrier at the interface was likely.

When the ion transferred is the salt of a weak acid, the transfer potential is pH-dependent. This has been observed by Sabela and co-workers [A36] for the transfer of nigericin across the water/nitrobenzene interface. The different equilibrium constants of this ionophore could be calculated from the pH dependence of the half-wave potential. Similar work has been carried out by Wang and Sun [A37] using bromocresol green, and the transferred dye was identified spectrophotometrically. Ion pairing with the organic electrolyte was apparent, as shown by the shift in $\Delta_o^w \phi_{1/2}$ with TBATPB concentration.

In classical electrochemical studies of the influence of specific adsorption of organic compounds on the rate of electron transfer, it has been recognized that the decrease in the rate constant can be due, among other reasons, to the extra work required to bring the reactant close to the metal surface. It has been known for some time that the adsorption of lecithin at the 1,2-DCE/water interface decreases the rate constant of ionic transfer. Recent work by Cunnane et al. [A38] has quantified the work of pore formation across a monolayer of adsorbed lecithin by relating the activation energy of transfer to the work of opening a pore against the surface pressure of the adsorbed film. TEA^+ was used as a probe ion, and the pore radius calculated was in remarkable agreement with the radius of the unhydrated ion. The energetics of transfer can therefore be related to the surface pressure of the monolayer.

Mediated Ion Transfer

The facilitated transfer of K^+ has been studied by Freiser and co-workers by polarography at the ascending water electrode and chronopotentiometry at stationary electrodes when the complexing agent was valinomycin [A39], dibenzo-18-cron-6 ether, DBC [A40,A41], and urushiol crown ether,

UCE [A41]. DBC and UCE gave two well-defined K^+ transfer waves, which were assigned to the transfer reactions:

$$CE(o) \rightleftharpoons CE(w) \tag{A1}$$

$$CE(w) + K^+(w) \rightleftharpoons KCE^+(w) \tag{A2}$$

$$KCE^+(w) \rightleftharpoons KCE^+(o) \tag{A3}$$

for low k^+ ion concentrations, and

$$K^+(w) + CE(o) \rightleftharpoons KCE^+(\sigma) \rightleftharpoons KCE^+(o) \tag{A4}$$

at high K^+ ion concentrations   (CE = crown ether, σ = interfacial region). The problem of establishing the site of the elementary rate determining step for crown ethers assisted interfacial ion transfers has also been investigated in detail by Kakutani and co-workers [A42] for the transfer of Na^+ ion facilitated by DBC at the nitrobenzene/water interface using ac impedance spectroscopy. The authors transposed the well-established reaction schemes of electron transfer in the presence of intermediate chemical steps that have been known for a long time for electrochemical reactions at metallic electrodes. Thus, the reaction sequence (A1), (A2), (A3) discussed by Yoshida and Freiser is entirely analogous to a classical CE mechanism; reaction (A4) corresponds to an E mechanism, whereas the equivalent to an EC reaction for potassium transfer would be

$$K^+(w) \rightleftharpoons K^+(o) \tag{A5}$$

followed by

$$K^+(o) + CE(o) \rightleftharpoons KCE^+(o) \tag{A6}$$

From the analysis of the ac impedance for Na^+ transfer, it was concluded that the transfer reaction in this case corresponded to an E mechanism. However, these results are not in disagreement with those of Yoshida and Freiser [A39-A41], since the analysis was restricted to the high Na^+ ion concentration region, where it has been proposed [A42] that reaction (A4) becomes rate-determining for the transfer of K^+. This E mechanism has recently been confirmed by Campbell and co-workers [A43] for the DCB-assisted transfer of K^+ at the water/1,2-DCE interface, using a micropipette electrode [A29] to measure the rate constants of transfer.

The possible application of electrochemical studies of assisted ion transfer across immiscible electrolytes to the important practical case of solvent extraction of metals has been studied by Homolka and Wendt [A44]. These authors studied the transfer potentials of Fe(II), Ni(II), and Zn(II) complexes with o-phenanthroline and o,o'-bypyridine and showed that complexation of Ni(II) by successive molecules of ligands changes ΔG° by an approximately equal amount of -37 kJ mol. Clear differences were noted in the behavior of the Zn(II) complexes owing to their lability and to disproportionation reactions in the organic phase. These studies are important in showing that, in principle, electrochemically assisted ion transfers can be used for metal separations. Another example of facilitated ion transfer of metal complexes is the work on the Cd(II)/2,2'-bipyridine system by Wang and Liu [A45], who observed the transfer of the 2:1 (ligand-to-cation) complex at the water-nitrobenzene interface. The specificity of the crown ethers in assisted ion transfers results from geometrical constraints for forming a suitable cage around the cation, in which hydration is replaced by the interaction with the electron lone pairs on the oxygen atoms. Yoshida and Kihara [A46] studied the effect of eliminating the rigid structure of a crown ether by replacing it with the flexible oxyethylene chains of Triton X. For Li^{+}, Na^{+}, NH_4^{+}, and K^{+}, the decrease in $\Delta G_{t,i}^{\circ}$ resulting from complexation was the same, indicating that the nature of the complex formed was very similar for these ions. Alkaline earth ions showed differences resulting from the very large differences in their degree of hydration. The assisted transfer of the H^{+} ion has also received some attention, and its transfer assisted by 18-crown-6 [A47] and by acridine chloride has been reported [A48].

Electron Transfer

The theory of electron transfer across liquid-liquid interfaces based on a pre-encounter equilibrium model has been presented by Girault and

Schiffrin [A49]. It was shown that the measurement of the rate constant at the potential of zero charge provides a simple link between experiment and theory, since the solvent reorganization energy (for outer sphere electron transfer) and the electrical work terms associated with bringing the reactants to the interface can be calculated from electrostatic models. There is a paucity of experimental results for this type of reaction, and one of the problems is the possibility of simultaneous electron and ion transfer occurring across the interface. This is particularly important for determining if the electron tunneling across the interface is occurring. For this reason, Geblewicz and Schiffrin [A50] studied the electron trans-fer to lutetium biphthalocyanine in 1,2-DCE from an aqueous hexacyano-ferrate couple, with a strongly salting-out aqueous medium, so that simultaneous ion and electron transfer were extremely unlikely to occur. In this case, the interfacial electron tunneling was observed. For fairly concentrated ferro-ferricyanide solutions, the aqueous phase behaves as a metallic electrode, and the rate constants for interfacial electron trans-fer could be related to those of the individual couples in contact with metallic electrodes. These ideas give a simple predictive route for the analysis of the important practical case of two phase redox reactions in synthetic organic chemistry, and this has been recently discussed [A51].

The transfer of ferricenium ion across the water-nitrobenzene inter-face has recently been studied by Hanzlik et al. [A52],who found that $\Delta_o^w \phi$ of transfer was ~100 mV more negative than that corresponding to the ferro-ferricenium (o)/ferro-ferricyanide (w) system. Thus, these results do not contradict a possible CEC reaction scheme:

$$Fc(o) \rightleftharpoons Fc(w) \tag{A7}$$

$$Fc(w) + [Fe(CN)_6]^{3-}(w) \rightleftharpoons Fc^+(w) + [Fe(CN)_6]^{4-}(w) \tag{A8}$$

$$Fc^+(w) \rightleftharpoons Fc^+(o) \tag{A9}$$

where Fc = ferrocene. Further work is necessary to elucidate this mechanism.

Analytical Application

These have utilized the property of polarized liquid-liquid interfaces to transfer ionic species selectively according to their Gibbs energies of transfer. This property is the main distinguishing feature of these systems, which has been put to use in many analytical applications. Wang and Liu [A53] have analyzed terramycin by measuring the ion transfer currents from the protonated form of the antibiotic and have obtained a detection limit of 5 g ml. Nitrate, perchlorate, and iodide were determined by Mareĉek et al. [A54] at a hanging electrolyte drop electrode. The support of the organic phase presents a considerable practical problem, and its incorporation into a gel is very advantageous. This method has been used for the determination of acetylcholine [A55] and perchlorate [A56] and for the construction of a voltammetric detector in a flow system [A57]. The study of the equilibrium physicochemical properties of liquid-liquid interfaces has also been applied recently to the understanding of the behavior of ion-selective electrodes [A58—A60]. Finally, the new technique of liquid microelectrodes [A29] has been used for the determination of acetylcholine, and it has been proposed that this method is suitable for monitoring biological fluids [A61].

REFERENCES TO APPENDIX

A1   J. D. Reid, O. R. Melroy, W. E. Bronner, P. Vanýsek, and R. P. Buck, In M. M. Kessler (Eds.), *Ion Measurements in Physiology and Medicine*, Springer-Verlag, Berlin, 1985.

A2   P. Vanýsek, *Electrochemistry on Liquid-Liquid Interfaces, Lecture Notes in Chemistry*, Vol.39, Springer-Verlag, New York, 1985.

A3   V. E. Kazarinov (Ed.), *The Interface Structure and Electrochemical Processes at the Boundary between Two Immiscible Electrolytes*, Springer-Verlag, Berlin, 1987.

A4   H. H. J. Girault, Electrochim. Acta 32:383—385 (1987).

A5   Z. Koczorowski, J. Electroanal. Chem. *190*:257—260 (1985).

A6   H. H. J. Girault and D. J. Schiffrin, Electrochim. Acta *31*:1341—1342 (1986).

A7   Z. Koczorowski, I. Paleska, and J. Kotowski, J. Electroanal. Chem. 235:287–298 (1987).

A8   Z. Samec, V. Mareĉek, K. Holub, S. Raĉinský, and P. Hájková, J. Electroanal. Chem. 225:65–78 (1987).

A9   G. M. Torrie and J. P. Valleau, J. Electroanal. Chem. 206:69–79 (1986).

A10   T. Osakai, T. Kakutani, and M. Senda, Bull. Chem. Soc. Jpn. 58:2626–2623 (1985).

A11   Z. Samec and V. Mareĉek, J. Electroanal. Chem. 200:17–33 (1986).

A12   T. Kakiuchi, M. Kobayashi, and M. Senda, Bull. Chem. Soc. Jpn. 60:3109–3115 (1987).

A13   D. J. Schiffrin, M. Wiles, and M. R. Calder, Electrochemical Society Proceedings, Vol 86-14, Electrochemical Sensors for Biomedical Applications, pp. 166–174.

A14   T. Wandlowski, S. Raĉinský, V. Mareĉek, and Z. Samec, J. Electroanal. Chem. 227:281–285 (1987).

A15   H. H. J. Girault and D. J. Schiffrin, Biochim. Biophys. Acta 857: 251–258 (1986).

A16   T. Makino and R. Aogaki, J. Electroanal. Chem. 198:209–212 (1986).

A17   I. Zagórska and Z. Koczorowski, J. Electroanal. Chem. 204:273–280 (1986).

A18   A. A. Kornyshev and A. G. Volkov, J. Electroanal. Chem. 180: 363–381 (1984).

A19   V. S. Markin and A. G. Volkov, J. Electroanal. Chem. 235:23–40 (1987).

A20   L. Cunningham and H. Freiser, Langmuir 1:537–541 (1985).

A21   G. Geblewicz, Z. Koczorowski, and Z. Figaszewski, Chemia Analityczna 31:567–575 (1986).

A22   Z. Pang, C. A. Chang, and E. Wang, J. Electroanal. Chem. 234: 71–84 (1987).

A23   H. Alemu and T. Solomon, J. Electroanal. Chem. 237:113–118 (1987).

A24   S. Kihara, M. Suzuki, K. Maeda, K. Ogura, S. Umetani, M. Matsui, and Z. Yoshida, Anal. Chem. 58:2954–2961 (1986).

A25   G. Geblewicz, A. K. Kontturi, K. Kontturi, and D. J. Schiffrin, J. Electroanal. Chem. 217:261–269 (1987).

A26   S. Kihara, M. Suzuki, K. Maeda, K. Ogura, and M. Matsui, J. Electroanal. Chem. 210:147–160 (1986).

A27   S. Kihara and Z. Yoshida, Talanta *31*:789–797 (1984).

A28   E. Wang and Z. Sun, Extended Abstracts, 169th Meeting of the Electrochemical Society, Boston, 1986, p. 857.

A29   G. Taylor and H. H. J. Girault, J. Electroanal. Chem. *208*:179–183 (1986).

A30   A. A. Stewart, J. Taylor, and H. H. J. Girault, J. Electroanal. Chem. (in press).

A31   J. Koryta, J. Electroanal. Chem. *213*:323–325 (1986).

A32   J. Koryta and M. Skalický, J. Electroanal. Chem. *229*:265–271 (1987).

A33   H. H. J. Girault and D. J. Schiffrin, J. Electroanal. Chem. *195:* 213–227 (1985).

A34   Y. Y. Gurevich and Y. I. Kharkats, J. Electroanal. Chem. *200:* 3–16 (1986).

A35   Z. Samec, Y. I. Kharkats, and Y. Y. Gurevich, J. Electroanal. Chem. *204*:257–266 (1986).

A36   A. Sabela, J. Koryta, and O. Valent, J. Electroanal. Chem. *204:* 267–272 (1986).

A37   E. Wang and Z. Sun, Anal. Chem. *59*:1414–1417 (1987).

A38   V. Cunnane, D. J. Schiffrin, M. Fleischmann, G. Geblewicz, and D. Williams, J. Electroanal. Chem. *243*:455–464 (1988).

A39   Z. Yoshida and H. Freiser, J. Electroanal. Chem. *179*:31–39 (1984).

A40   L. Sinru and H. Freiser, J. Electroanal. Chem. *191*:437–439 (1985).

A41   L. Sinru, Z. Zaofan, and H. Freiser, J. Electroanal. Chem. *210:* 137–146 (1986).

A42   T. Kakutani, Y. Nishiwaki, T. Osakai, and M. Senda, *Bull. Chem. Soc. Jpn.* *59*:781–788 (1986).

A43   J. A. Campbell, A. A. Stewart, and H. H. J. Girault, J. Chem. Soc. Faraday Trans. I (in press).

A44   D. Homolka and H. Wendt, Ber. Bunsenges. Phys. Chem. *89*:1075–1082 (1985).

A45   E. Wang and Y. Liu, J. Electroanal. Chem. *214*:465–472 (1986).

A46   Z. Yoshida and S. Kihara, J. Electroanal. Chem. *227*:171–181 (1987).

A47   E. Makrlik, A. Hofmanova, and L. Q. Hung, J. Colloid Interface Sci. *107*:1–4 (1985).

A48   Y. Liu and E. Wang, J. Electroanal. Chem. *234*:85–92 (1987).

A49 H. H. J. Girault and D. J. Schiffrin, J. Electroanal. Chem. *244:* 15–26 (1988).

A50 G. Geblewicz and D. J. Schiffrin, J. Electroanal. Chem. *244:*27–37 (1988).

A51 V. Cunnane, D. J. Schiffrin, C. Beltran, G. Geblewicz, and T. Solomon, J. Electroanal. Chem. (in press).

A52 J. Hanzlík, Z. Samec, and J. Hovorka, J. Electroanal. Chem. *216:* 303–308 (1987).

A53 E. Wang and Y. Liu, J. Electroanal. Chem. *214:*459–464 (1986).

A54 V. Mareĉek, H. Jänchenová, Z. Samec, and M. Bĩezina, Analytica Chimica Acta *185:*359–362 (1986).

A55 T. Kakutani, T. Ohkouchi, T. Osakai, T. Kakiuchi, and M. Senda, Analytical Sciences *1:*219–225 (1985).

A56 B. Hundhammer, S. K. Dhawan, A. Bekele, and H. J. Seidlitz, J. Electroanal. Chem. *217:*253–259 (1987).

A57 V. Mareĉek, H. Jänchenová, M. P. Colombini, and P. Papoff, J. Electroanal. Chem. *217:*213–219 (1987).

A58 T. Kakiuchi, I. Obi, and M. Senda, Bull. Chem. Soc. Jpn. *58:* 1636–1641 (1985).

A59 B. Hundhammer, H. J. Seidlitz, S. Becker, and S. K. Dhawan, J. Electroanal. Chem. *180:*355–362 (1984).

A60 T. Kakiuchi and M. Senda, Bull. Chem. Soc. Jpn. *60:*3099–3107 (1987).

A61 M. Senda, T. Kakutani, T. Osakai, and T. Ohkouchi, in E. Pungor (Ed.), *Bioelectroanalysis 1*, Akademiai Kiado, Budapest, 1987, pp. 353–361.

ACKNOWLEDGMENTS

The authors wish to thank the Institut National Polytechnique de Grenoble (France) and the Direction des Relations Extérieures du C.N.R. S. (France) for financial support during the course of this work. Part of this work was supported by the S.E.R.C. (England). Helpful discussions and comments by Professor R. Parsons are gratefully acknowledged.

REFERENCES

1. W. Nernst and E. H. Riesenfield, Ann. Phys. *8*:600 (1902).

2. W. Nernst, Z. Phys. Chem. *2*:613 (1888).

3. H. J. S. Sand, Philos. Mag. *1*:45 (1900).

4. E. H. Riesenfield, Ann. Phys. *8*:609 (1902).

5. E. H. Riesenfield, Ann. Phys. *8*:616 (1902).

6. E. H. Riesenfield and B. Reinhold, Z. Phys. Chem. *68*:459 (1909).

7. M. Cremer, Z. Biol. *47*:562 (1906).

8. W. Ostwald, Z. Phys. Chem. *6*:71 (1890).

9. R. Beutner, J. Am. Chem. Soc. *35*:344 (1913).

10. E. Baur and S. Kronman, Z. Phys. Chem. *92*:81 (1917).

11. R. Beutner, Trans. Am. Electrochem. Soc. *21*:219 (1912).

12. R. Beutner, Z. Elektrochem. *19*:319 (1913).

13. R. Beutner, Z. Elektrochem. *19*:467 (1913).

14. R. Beutner, Z. Elektrochem. *24*:94 (1918).

15. R. Beutner, Z. Phys. Chem. *104*:472 (1923).

16. E. Baur, Z. Elektrochem. *24*:100 (1918).

17. E. Baur, Z. Elektrochem. *28*:421 (1922).

18. E. Baur, Z. Phys. Chem. *103*:39 (1922).

19. E. Baur, Z. Phys. Chem. *106*:157 (1923).

20. E. Baur and E. Alleman, Z. Elektrochem. *32*:574 (1926).

21. G. Ehrensvärd and D. F. Cheesman, Science *94*:23 (1941).

22. G. Ehrensvärd and D. F. Cheesman, Sven. Kem. Tidskr. *53*:126 (1941).

23. G. Ehrensvärd and L. G. Sillen, Nature *141*:788 (1938).

24. G. Ehrensvärd and L. G. Sillen, Nature *142*:396 (1938).

25. G. Ehrensvärd and L. G. Sillen, Z. Elektrochem. *45*:440 (1939).

26. W. Wilbrandt, Ergeb, Physiol. *40*:204 (1938).

27. S. Craxford, O. Gatty, and Lord Rothschild, Nature *141*:1098 (1938).

28. R. B. Dean, O. Gatty, and E. K. Rideal, Trans. Faraday Soc. *36:* 161 (1940).

29. R. B. Dean, Trans. Faraday Soc. 36166 (1940).

30. F. M. Karpfen and J. E. B. Randles, Trans. Faraday Soc. *49*:823 (1953).

31. K. F. Bonhoeffer, M. Kahlweit, and H. Strehlow, Z. Elektrochem. *57*:614 (1953).

32. T. Teorell, Proc. Soc. Exp. Biol, *33*:282 (1935).

33. K. H. Meyer, H. Hauptmann, and J. F. Sievers, Helv. Chim. Acta *19*:948 (1936).

34. K. F. Bonhoeffer, M. Kahlweit, and H. Strehlow, Z. Phys. Chem. *1*:21 (1954).

35. M. Kahlweit, H. Strehlow, and C. S. Hocking, Z. Phys. Chem. *4*: 212 (1955).

36. J. T. Davies and E. K. Rideal, Can. J. Chem. *33*:947 (1955).

37. J. T. Davies, Z. Elektrochem. *55*:559 (1951).

38. J. T. Davies, Proc. 3rd Int. Congr. Surf. Act., Cologne *2*:254 (1960).

39. M. Dupeyrat, J. Chim. Phys. *61*:306 (1964).

40. M. Dupeyrat, J. Chim. Phys. *61*:323 (1964).

41. C. Gavach, J. Chim. Phys. *64*:799 (1967).

42. C. Gavach, J. Chim. Phys. *64*:810 (1967).

43. E. J. W. Verwey and K. F. Niessen, Philos. Mag. *28*:435 (1939).

44. M. Kahlweit and H. Strehlow, Z. Elektrochem. *58*:658 (1954).

45. H. Strehlow, Z. Elektrochem. *59*:744 (1955).

46. H. D. Ohlenbusch, Z. Elektrochem. *60*:607 (1956).

47. J. Guastalla, J. Chim. Phys. *53*:470 (1956).

48. J. Guastalla, Proc. 2nd Int. Congr. Surf. Act. *3*:112 (1957).

49. A. Watanabe, M. Matsumoto, and R. Gotoh, Bull. Inst. Chem. Res. Kyoto Univ. *44*:273 (1966).

50. A. Watanabe, M. Matsumoto, H. Tamai, and R. Gotoh, Kolloid Z. Z. Polym. *220*:152 (1967).

51. A. Watanabe, M. Matsumoto, H. Tamai, and R. Gotoh, Kolloid Z. Z. Polym. *221*:47 (1967).

52. A. Watanabe, M. Matsumoto, H. Tamai, and R. Gotoh, Kolloid Z. Z. Polym. *228*:58 (1968).

53. A. Watanabe, A. Fukii, Y. Sakamori, K. Higashitsuji, and H. Tamai, Kolloid Z. Z. Polym. *243*:42 (1971).

54. A. Watanabe and H. Tamai, Kolloid Z. Z. Polym. *246*:587 (1971).

55. A. Watanabe, H. Tamai, and K. Higashitsuji, J. Colloid Interface Sci. *43*:548 (1973).

56. A. Watanabe and H. Tamai, Hyomen *13*:67 (1975).

57. M. Blank and S. Feig, Science *141*:1173 (1963).

58. M. Blank, Proc. 4th Int. Congr. Surf. Act. Substs., Brussels *2*:233 (1964).

59. M. Blank, J. Colloid Interface Sci. *22*:51 (1966).

60. M. Dupeyrat and J. Michel, C. R. Acad. Sci. *C264*:1240 (1967).

61. M. Dupeyrat and J. Michel, J. Colloid Interface Sci. *29*:605 (1969).

62. M. Dupeyrat and J. Michel, Experientia Suppl. *18*:269 (1971).

63. M. Dupeyrat and E. Nakache, J. Colloid Interface Sci. *73*:332 (1980).

64. C. Gavach and B. D'Epenoux, C. R. Acad. Sci. *C272*:872 (1971).

65. C. Gavach, Experimentia Suppl. *18*:321 (1971).

66. P. Joos and M. Van Bockstacle, J. Phys. Chem. *80*:1573 (1976).

67. Y. Verburgh and P. Joos, J. Colloid Interface Sci. *74*:384 (1980).

68. C. Gavach, T. Mlodnicka, and J. Guastalla, C. R. Acad. Sci. *C266*: 1196 (1968).

69. J. Guastalla, C. R. Acad. Sci. *C269*:1360 (1969).

70. J. Guastalla, Nature *227*:485 (1970).

71. J. Guastalla, Experimentia Suppl. *18*:333 (1971).

72. J. Guastalla and G. Bertrand, C. R. Acad. Sci. *C274*:1884 (1972).

73. J. Guastalla and C. Bertrand, C. R. Acad. Sci. *C277*:279 (1973).

74. J. W. Gibbs, *Collected Works* (2nd Ed.), Longman, New York, Vol. 1, p. 429.

75. E. A. Guggenheim, J. Phys. Chem. *33*:842 (1929).

76. D. Bauer and M. Breant, Electroanal. Chem. *8*:282 (1975).

77. A. J. Parker, Chem. Rev. *69*:1 (1969).

78. V. A. Pleskov, Usp. Khim. *16*:254 (1947).

79. H. M. Koepp, H. Wendt, and H. Strehlow, Z. Elektrochem. *64*:483 (1960).

80. R. Alexander and A. J. Parker, J. Am. Chem. Soc. *90*:3313 (1968).

81. E. Grunwald, G. Baughman, and G. Kohnstam, J. Am. Chem. Soc. *82*:5801 (1960).

82. A. J. Parker, J. Chem. Soc. A. *1966*:220.

83. R. Parsons and B. T. Rubin, J. Chem. Soc. Faraday Trans. 1 70:1636 (1974).

84. M. H. Abraham and J. Liszi, J. Inorg. Nucl. Chem. 43:143 (1981).

85. Y. Marcus, Pure Appl. Chem. 55:977 (1983).

86. M. H. Abraham and J. Liszi, J. Chem. Soc. Faraday Trans. 1 74: 1604 (1978).

87. M. H. Abraham and J. Liszi, J. Chem. Soc. Faraday Trans. 1 74: 2858 (1978).

88. M. H. Abraham, J. Liszi, and L. Meszaros, J. Chem. Phys. 70:2491 (1979).

89. M. H. Abraham and J. Liszi, J. Chem. Soc. Faraday Trans. 1 76: 1219 (1980).

90. M. H. Abraham, J. Liszi, and E. Papp, J. Chem. Soc. Faraday Trans. 1 78:197 (1982).

91. J. Czapkiewicz and B. Czapkiewicz-Tutaj, J. Chem. Soc. Faraday Trans. 1 76:1663 (1980).

92. J. P. Antonie, I. de Aguirre, F. Janssens, and F. Thyrion, Bull. Soc. Chim. Fr. 2 1980:207.

93. M. H. Abraham and A. F. Danil de Namor, J. Chem. Soc. Faraday Trans. 1 72:955 (1976).

94. J. Rais, Collect. Czech. Chem. Commun. 36:3253 (1971).

95. J. Rais, P. Selucky, and A. Kyrs, J. Inorg. Nucl. Chem. 38:1376 (1976).

96. A. F. Danil de Namor and T. Hill, J. Chem. Soc. Faraday Trans. 1 79:2713 (1983).

97. V. Marecek and Z. Samec, Anal. Chim. Acta 151:265 (1983).

98. C. Gavach and F. Henry, J. Electroanal. Chem. 54:361 (1974).

99. M. Gros, S. Gromb, and C. Gavach, J. Electroanal. Chem. 89:29 (1978).

100. C. Gavach and N. Davion, Electrochim. Acta 18:649 (1973).

101. P. Vanysek and M. Behrendt, J. Electroanal. Chem. 130:287 (1981).

102. B. Vanysek, J. Electroanal. Chem. 121:149 (1981).

103. M. Gerin and J. Fresco, Anal. Chim. Acta 97:165 (1978).

104. J. Koryta, Electrochim. Acta 29:445 (1984).

105. Z. Koczorowski, J. Electroanal. Chem. 127:11 (1981).

106. M. I. Gugeshashvili, M. A. Manvelyan, and L. I. Boguslavsky, Sov. Electrochem. 10:782 (1974).

107. Z. Koczorowski and G. Geblewicz, J. Electroanal. Chem. *152*:55 (1983).

108. Z. Koczorowski and S. Minc, Electrochim. Acta *8*:645 (1963).

109. S. Minc and Z. Koczorowski, Electrochim. Acta *8*:575 (1963).

110. L. I. Boguslavsky and M. I. Gugeshashvili, Sov. Electrochem. *8*:1433 (1972).

111. M. I. Gugeshashvili and L. I. Boguslavsky, Sov. Electrochem. *9*:661 (1973).

112. M. I. Gugeshashvili, B. T. Lozhkin, and L. I. Boguslavsky, Sov. Electrochem. *10*:1218 (1974).

113. M. A. Manvelyan, V. N. Andreev, V. E. Kazarinov, and L. I. Boguslavsky, Sov. Electrochem. *13*:1531 (1977).

114. O. R. Melroy and R. P. Buck, J. Electroanal. Chem. *143*:23 (1983).

115. L. Q. Hung, J. Electroanal. Chem. *115*:159 (1980).

116. L. Q. Hung, J. Electroanal. Chem. *149*:1 (1983).

117. F. G. Donnan, Chem. Rev. *1*:73 (1924).

118. P. Joos, L. Jansegers, and Y. Verburch, J. Colloid Interface Sci. *63*:27 (1978).

119. P. Joos, Y. Verburgh, Bull. Soc. Chim. Belg. *87*:737 (1978).

120. D. A. McInnes, *The Principles of Electrochemistry*, Dover, New York, 1961.

121. R. Bennes and B. E. Conway, Can. J. Chem. *59*:1978 (1981).

122. Z. Koczorowski and I. Zagorska, J. Electroanal. Chem. *159*:183 (1983).

123. R. Parsons, Can. J. Chem. *37*:308 (1959).

124. R. S. Hansen, J. Phys. Chem. *66*:410 (1962).

125. R. Defay, I. Prigogine, A. Bellemans, and D. H. Everett, *Surface Tension and Adsorption*, Longmans, London, 1966.

126. L. I. Boguslavsky, A. N. Frumkin, and M. I. Gugeshashvili, Sov. Electrochem. *12*:799 (1976).

127. M. A. Manvelyan, G. L. Neugkdova, and L. I. Boguslavsky, Sov. Electrochem. *12*:1145 (1976).

128. C. Gavach, H. Bellet, N. Kavion, and P. Seta, J. Chim. Phys. *68*:1005 (1971).

129. C. Gavach, P. Seta, and B. D'Epenoux, J. Electroanal. Chem. *83*: 225 (1977).

130. H. H. Girault and D. J. Schiffrin, J. Electroanal. Chem. *150*:43 (1983).

131. S. Trasatti and R. Parsons, Pure Appl. Chem. 55:1252 (1983).

132a. R. Aveyard, S. M. Saleem, and R. Heselden, J. Chem. Soc. Faraday Trans. 1 73:84 (1977).

132b. R. Aveyard and S. M. Saleem, J. Chem. Soc. Faraday Trans. 1 73:896 (1977).

133. R. Aveyard and B. Vincent, Prog. Surf. Sci. 8:59 (1977).

134. A. N. Frumkin and A. Slygin, Acta Physicochim. URSS 5:819 (1937).

135. R. Parsons, Adv. Electrochem. Electrochem. Eng. 7:177 (1970).

136. T. Kakiuchi and M. Senda, Bull. Chem. Soc. Jpn. 56:1322 (1983).

137. T. Kakiuchi and M. Senda, Bull. Chem. Soc. Jpn. 56:1753 (1983).

138. T. Kakiuchi and M. Senda, Bull. Chem. Soc. Jpn. 56:2912 (1983).

139. H. H. Girault and D. J. Schiffrin, J. Electroanal. Chem. 170:127 (1984).

140. Z. Samec, V. Marecek, and D. Homolka, J. Electroanal. Chem. 126:121 (1981).

141. V. Marecek and Z. Samec, J. Electroanal. Chem. 149:185 (1983).

142. P. Hajkova, D. Homolka, V. Marecek, and Z. Samec, J. Electroanal. Chem. 151:277 (1983).

143. D. Homolka, P. Hajkova, V. Marecek, and Z. Samec, J. Electroanal. Chem. 159:233 (1983).

144. Z. Samec, V. Marecek, and D. Homolka, Faraday Discuss. Chem. Soc. 77:277 (1984).

145. Z. Samec, V. Marecek, and D. Homolka, J. Electroanal. Chem. 170:383 (1984).

146. V. Marecek and Z. Samec, J. Electroanal. Chem. 185:263 (1985).

147. Z. Samec, V. Marecek, and D. Homolka, J. Electroanal. Chem. 187:31 (1985).

148. A. N. Frumkin, N. A. Balashova, and V. E. Kazarinov, J. Electrochem. Soc. 113:1011 (1966).

149. H. H. Girault and D. J. Schiffrin, J. Electroanal. Chem. 161:415 (1984).

150. J. D. Reid, P. Vanysek, and R. P. Buck, J. Electroanal. Chem. 161:1 (1984).

151. J. Reid, P. Vanysek, and R. Buck, J. Electroanal. Chem. 170:109 (1984).

152. C. Gavach and F. Henry, C. R. Acad. Sci. C274:1545 (1972).

153. C. Gavach, J. Chim. Phys. *701478* (1973).

154. C. Gavach, F. Henry, and R. Sandeaux, C. R. Acad. Sci. *C278*: 491 (1974).

155. C. Gavach, P. Seta, and F. Henry, Bioelectrochem. Bioenerg. *1*: 329 (1974).

156. C. Gavach and B. D'Epenoux, J. Electroanal. Chem. *55*:59 (1974).

157. C. Gavach, B. D'Epenoux, and F. Henry, J. Electroanal. Chem. *64*:107 (1975).

158. B. D'Epenoux, P. Seta, G. Amblard, and C. Gavach, J. Electroanal. Chem. *99*:77 (1979).

159. Z. Samec, J. Electroanal. Chem. *99*:197 (1979).

160. Z. Samec, V. Marecek, J. Koryta, and M. W. Khalil, J. Electroanal. Chem. *83*:393 (1977).

161. Z. Samec, V. Marecek, and J. Weber, J. Electroanal. Chem. *100*: 841 (1979).

162. Z. Samec, V. Marecek, J. Weber, and D. Homolka, J. Electroanal. Chem. *126*:105 (1981).

163. Z. Samec, J. Electroanal. Chem. *111*:211 (1980).

164. O. R. Melroy, R. P. Buck, F. S. Stover, and H. C. Hughes, J. Electroanal. Chem. *121*:93 (1981).

165. O. R. Melroy, W. E. Bronner, and R. P. Buck, J. Electrochem. Soc. *130*:373 (1983).

166. O. R. Melroy and R. P. Buck, J. Electroanal. Chem. *136*:19 (1982). Errata, J. Electroanal. Chem. *147*:351 (1983).

167. O. R. Melroy and R. P. Buck, J. Electroanal. Chem. *151*:1 (1983).

168. W. E. Bronner, O. R. Melroy, and R. P. Buck, J. Electroanal. Chem. *162*:263 (1984).

169. T. Osakai, T. Kakutani, and M. Senda, Bull. Chem. Soc. Jpn. *57*:370 (1984).

170. H. H. Girault and D. J. Schiffrin, J. Electroanal. Chem. (in press)

171. A. E. Stearn and H. Eyring, J. Phys. Chem. *44*:955 (1940).

172. A. J. Bard and L. R. Faulkner, *Electrochemical Methods: Fundamentals and Applications,* Wiley, New York, 1980.

173. Southampton Electrochemistry Group, *Instrumental Methods for Electrochemistry,* Ellis Horwood, Chichester, West Sussex, England, 1985.

174. P. Seta and C. Gavach, C. R. Acad. Sci. *C275*:1231 (1972).

175. P. Seta and C. Gavach, C. R. Acad. Sci. *C277*:403 (1973).

176. D. Schuhman and P. Seta, J. Colloid Interface Sci. *69*:448 (1979).

177. D. Schuhman and P. Seta, Physicochem. Hydrodyn. *1*:57 (1980).

178. J. Koryta, P. Vanysek, and M. Brezina, J. Electroanal. Chem. *67*:263 (1976).

179. J. Koryta, P. Vanysek, and M. Brezina, J. Electroanal. Chem. *75*:211 (1977).

180. J. Koryta, Electrochim. Acta *24*:293 (1979).

181. Z. Figazewski, Z. Koczorowski, and G. Geblewicz, J. Electroanal. Chem. *139*:317 (1982).

182. Z. Samec, V. Marecek, J. Weber, and D. Homolka, J. Electroanal. Chem. *99*:385 (1979).

183. T. Kakutani, T. Osakai, and M. Senda, Bull. Chem. Soc. Jpn. *56*:991 (1983).

184. B. Hundhammer, T. Solomon, and H. Alemu, J. Electroanal. Chem. *149*:179 (1983).

185. L. A. Berlouis, J. Chatman, H. H. Girault, and D. J. Schiffrin, *Extended Abstracts*, 163rd Meet. Am. Electrochem. Soc., San Francisco, May 1983.

186. S. Kihara, Z. Yoshida, and T. Fujinaga, Bunseki Kagaku *31*:E297 (1982); *31*:E301 (1982).

187. T. Osakai, T. Kakutani, Y. Nishiwaki, and M. Senda, Bunseki Kagaku *32*:E81 (1983).

188. B. Hundhammer and T. Solomon, J. Electroanal. Chem. *157*:19 (1983).

189. Z. Samec, D. Homolka, V. Marecek, and L. Kavan, J. Electroanal. Chem. *145*:213 (1983).

190. D. Homolka and V. Marecek, J. Electroanal. Chem. *112*:91 (1980).

191. B. Hundhammer, T. Solomon, and B. Alemayehu, J. Electroanal. Chem. *135*:301 (1982).

192. Z. Koczorowski and G. Geblewicz, J. Electroanal. Chem. *108*:117 (1980).

193. Z. Koczorowski and G. Geblewicz, J. Electroanal. Chem. *139*:177 (1982).

194. G. Geblewicz, Z. Koczorowski, and Z. Figaszewski, Colloids Surf. *6*:43 (1983).

195. G. Geblewicz, Z. Figaszewski, and Z. Koczorowski, J. Electroanal. Chem. *177*:1 (1984).

196. Z. Koczorowski, G. Geblewicz, and I. Paleska, J. Electroanal.
      Chem. *172*:327 (1984).

197. T. Solomon, H. Alemu, and B. Hundhammer, J. Electroanal. Chem.
      169:303 (1984).

198. Z. Koczorowski, I. Paleska, and G. Geblewicz, J. Electroanal.
      Chem. *164*:201 (1984).

199. T. Solomon, H. Alemu, and B. Hundhammer, J. Electroanal. Chem.
      169:311 (1984).

200. D. Homolka, L. Q. Hung, A. Hofmanova, M. W. Kahlil, J. Koryta,
      V. Marecek, Z. Samec, S. K. Sen, P. Vansek, J. Weber, M. Bre-
      zina, M. Janda, and I. Stebor, Anal. Chem. *52*1606 (1980).

201. A. Hofmanova, L. Q. Hung, and M. W. Kahlil, J. Electroanal.
      Chem. *135*:257 (1982).

202. P. Vanysek, W. Ruth, and J. Koryta, J. Electroanal. Chem. *148:*
      117 (1983).

203. E. Makrlik, L. Q. Hung, and A. Hofmanova, Electrochim. Acta
      28:847 (1983).

204. J. Koryta, Electrochim. Acta *29*:445 (1984).

205. D. Homolka, K. Holub, and V. Marecek, J. Electroanal. Chem.
      138:29 (1982).

206. D. Homolka, V. Marecek, Z. Samec, O. Ryba, and J. Petranek,
      J. Electroanal. Chem. *125*:243 (1981).

207. Z. Samec, D. Homolka, and V. Marecek, J. Electroanal. Chem.
      135:265 (1982).

208. J. Koryta, W. Ruth, P. Vanysek, and A. Hofmanova, Anal. Lett.
      B15:1685 (1982).

209. G. Du, J. Koryta, W. Ruth, and P. Vanysek, J. Electroanal.
      Chem. *159*:413 (1983).

210. E. Makrlik, W. Ruth, and P. Vanysek, Electrochim. Acta *28*:575
      (1983).

211. D. Homolka, V. Marecek and Z. Samec, K. Base, and H. Wendt,
      J. Electroanal. Chem. *163*:159 (1984).

212. Z. Yoshida and H. Freiser, J. Electroanal. Chem. *162*:307 (1984).

213. Ju. N. Kozlov and J. Koryta, Anal. Lett. *B16*:255 (1983).

214. V. Marecek and Z. Samec, Anal. Chim. Acta *141*:65 (1982).

215. J. D. Reid, O. R. Melroy, and R. P. Buck, J. Electroanal. Chem.
      147:71 (1983).

216. V. Marecek and Z. Samec, Anal. Lett. *14(B15)*:1241 (1981).

217. B. Douglas, D. H. McDaniel, and J. J. Alexander, *Concepts and Models of Inorganic Chemistry*, Wiley, New York, 1983.

218. R. A. Marcus, Ann. Rev. Phys. Chem. *15*:155 (1964).

219. H. H. J. Girault and D. J. Schiffrin, in *Charge and Field Effects in Biosystems* (M. J. Allen and P. N. R. Usherwood, eds.), Abacus Press, Turnbridge Wells, Kent, England, 1984, pp. 171–178.

220. H. H. J. Girault and D. J. Schiffrin, J. Electroanal. Chem. *179:* 277 (1984).

221. M. Guainazzi, G. Silvestri, and G. Survalle, J. Chem. Soc. D. 200 (1975).

222. Z. Samec, V. Marecek, and J. Weber, J. Electroanal. Chem. *103:* 11 (1979).

223. H. Gerischer, Adv. Electrochem. Electrochem. Eng. *1*:139 (1961).

224. J. I. Kharkats and A. G. Volkov, J. Electroanal. Chem. *184*:435 (1985).

225. H. H. J. Girault and D. J. Schiffrin, J. Electroanal. Chem. (submitted).

226. N. Sutin, Acc. Chem. Res. *15*:275 (1982).

ELLIPSOMETRY: PRINCIPLES AND
RECENT APPLICATIONS IN ELECTROCHEMISTRY

Shimshon Gottesfeld

Los Alamos National Laboratory
Los Alamos, New Mexico

I. INTRODUCTION

Ellipsometry is a technique used to obtain the dielectric properties of
materials in the optical domain. The technique is based on the measurement of changes in the state of polarization of a collimated monochromatic
light beam caused by the interaction of the beam with the physical system investigated. Ellipsometry has usually been employed in a reflection
mode. In this mode, information on the optical properties of a reflecting
interface is derived from the measured relationship between the states
of polarization of a well-defined incident beam and the resulting specularly reflected beam. Ellipsometry has been used quite extensively in
the reflection mode for the determination of the optical spectra of highly
absorbing solids, including metal and semiconductor phases. However,
its widest use has been in the optical characterization of thin films.
This use has also been the main source of interest in ellipsometry as a
tool in interfacial electrochemistry, an area in which it has usually been
employed within the near-infrared (IR), visible, and near-ultraviolet
(UV) spectral regions. The attractive features of ellipsometry as a
technique for the examination of electrochemical interfaces are the
following:

1.   Ellipsometric mesurements are performed in situ. Changes in the
     optical properties of the electrochemical interface can be followed
     with the incident and reflected beams passing through the liquid

electrolyte. The electrode is thus always examined while immersed in the electrolyte and under potential control.

2. Ellipsometry exhibits high sensitivity to changes in surface coverage. For example, detection of the electrochemical growth of 0.1% of a monolayer of an oxide film on a metal substrate is well within the capability of a standard instrument. Lower detection levels are usually prevented not by instrumentation, but by the limited purity and/or stability of real electrochemical systems.

3. Ellipsometry is highly surface selective. For example, in a case of formation of both a surface film and an absorbing product in solution, the ellipsometric parameters will respond only to the surface process.

4. Compared with transmission photometry for examining surface films on electrodes, ellipsometry allows the determination of optical properties and their spectra for films formed on regular bulk metal or semiconductor electrodes, while transmission photometry requires thin-film electrodes deposited on transparent substrates. Furthermore, ellipsometry provides more information on the examined surface film. Both the real and imaginary components of the complex dielectric constant (or complex refractive index) and the thickness of the film can be derived from the results of ellipsometric measurements. Also, recent spectroellipsometric measurements have proved to be an effective source of information on the microstructure of composite surface films.

5. With the recent development of various modes of automatic ellipsometry, transient phenomena can be followed with time resolution of 1 msec or better. Thus the dynamics of a large number of interfacial processes can be followed optically in situ with automated ellipsometers.

Despite this list of apparently attractive features, there have been several obstacles to the wider and better use of ellipsometry by electrochemists. One of the obstacles is that ellipsometry is somewhat less

straightforward than other spectroscopic techniques. The information,
in the form of optical properties at a single wavelength or their spectrum,
is not obtained directly and has to be extracted from the raw results
with the aid of a computer program. However, the requirement of a
computer as an integral part of the experimental system has become very
easy to fulfill in the era of the microprocessor. Another apparent reason
for caution has been the relatively extensive use of ellipsometry in the
sixties and early seventies for the investigation of processes on electrode
surfaces on the submonomolecular level, e.g., potential-modulated electrosorption of ions and of H and O atoms. As discussed below, the small
ellipsometric signals obtained during submonolayer electrosorption are
due to simultaneous contributions from several interfacial phenomena,
and these signals are relatively difficult to interpret quantitatively.
The complexity of interpretation in these specific cases seems to have
left an unjustified impression that the analysis of ellipsometric results
obtained for any electrochemical system may be too complex. It will be
shown below that in investigations of many electrochemical systems,
ellipsometric measurements supply the complete information required for
the unambiguous optical characterization of a surface film. Furthermore,
the optical properties of the film in the visible and near UV will be
shown to supply valuable information, such as film thickness, film density and internal morphology, and the film's electronic properties. The
ellipsometric data are particularly informative when measurements are
taken over a wide range of wavelengths ("spectroscopic ellipsometry").
It will also be shown that time-resolved ellipsometry can provide information on the dynamics of interfacial processes at the metal-electrolyte
interface, as well as the dynamics of processes within electroactive films
associated with redox reactivity or with an insulator-conductor transition.
Ellipsometry is indeed a natural tool in the field of surface-modified or
filmed electrodes, which have recently become a central subject of
investigation in electrochemistry.

A book [1] and several reviews on the theory and applications of ellipsometry have appeared during the last 10 years. The reader is referred also to Proceedings of the International Conference on Ellipsometry, the most recent of which meetings took place in 1979 and in 1983 [2,3]. Several reviews on the use of ellipsometry specifically for electrochemistry have also been published [4–7]. The purpose of this chapter is to review recent applications of ellipsometry in electrochemistry. The preceding discussion of principles, instrumentation, and the general nature of information obtained is intended to allow readers to use this chapter also as an introduction to the application of ellipsometry in electrochemistry. This chapter does not, however, include a comprehensive discussion of the physics of polarized light. It is felt that the references cited above cover this subject and can be consulted, when required for further details.

The following section describes basic parameters that determine the properties of a light beam specularly reflected from a film-free electrode surface and from a filmed electrode surface immersed in a liquid electrolyte. This section also describes equations by which these parameters are related. Section III deals with the principles of the instrumental techniques employed for the automatic measurement of ellipsometric parameters and gives some details on the instrumental systems most widely employed. Section IV considers the range of optical properties and the nature of optical spectra encountered in films formed in electrochemical interfaces and describes the qualitative trends of ellipsometric effects brought about by such films. Section V describes recent experimental results obtained in ellipsometric investigations of electrochemical systems, including investigations of the growth, the optical properties, and electrochemical conversion processes in oxide and in polymeric films (Sec. V.A), ellipsometric measurements of thin chemisorbed films and double-layer effects (Sec. V.B), of composite surface films (Sec. V.C), of UPD metal layers and thicker metal electrodeposits (Sec. V.D), of

corrosion and passivation processes (Sec. V.E), and of surface layers
on semiconductor electrodes (Sec. V.F).

II.   SPECULAR REFLECTION OF LIGHT FROM
THE SOLID-ELECTROLYTE INTERFACE

A.   Fresnel Reflection Coefficients for a
Film-Free and for a Filmed Substrate

Ellipsometric measurements in electrochemical systems usually begin with
a film-free substrate, which has to be characterized optically first.
This is followed by the electrochemical growth and ellipsometric characterization of the surface film to be studied.   The states of the film-free
and the filmed substrates are shown schematically in Fig. 1.   In the
equations below the semi-infinite ambient phase (electrolyte solution)
is designated by 1, the surface film by 2, and the semi-infinite metal
(or semiconductor) electrode substrate by 3.   A collimated monochromatic beam incident at an angle of incidence ϕ_1 on the bare substrate is
specularly reflected at the same angle ϕ_1 (only specular reflection is

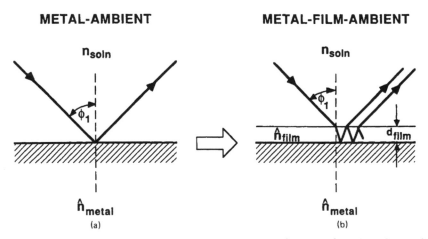

FIG. 1.   Schematic presentation of specular reflection for the electrode-
electrolyte interface: (a) in the absence of any surface film; (b) in the
presence of a surface film.

considered in this section). The ratio, r, between the electric field vectors in the reflected beam, $E^{(r)}$, and in the incident beam, $E^{(in)}$, is the basic parameter that provides information on the optical characteristics of the reflecting interface. This ratio, which is named the Fresnel reflection coefficient, depends on the state of polarization of the incident beam. For light polarized parallel to the plane of incidence (p, polarization) or perpendicular to the plane of incidence (s, polarization), the reflection coefficients are thus defined as

$$\hat{r}_p = \frac{E_p^{(r)}}{E_p^{(in)}} \qquad \hat{r}_s = \frac{E_s^{(r)}}{E_s^{(in)}} \tag{1}$$

For a metal-ambient system, i.e., a system of two adjacent semi-infinite phases [Fig. 1(a)], the reflection coefficients are given in terms of the optical properties of the two phases and the angle of incidence, ϕ_1, by [4,6]

$$\hat{r}_p^{13} = \frac{n_1 \cos \phi_3 - \hat{n}_3 \cos \phi_1}{n_1 \cos \phi_3 + \hat{n}_3 \cos \phi_1} \tag{2a}$$

and

$$\hat{r}_s^{13} = \frac{\hat{n}_3 \cos \phi_3 - n_1 \cos \phi_1}{n_1 \cos \phi_1 + \hat{n}_3 \cos \phi_3} \tag{2b}$$

where n_1 and $\hat{n}_3$ are the refractive indices of the electrolyte and the metal, respectively, and ϕ_3, the angle of propagation within the metal phase, is defined according to Snell's law,

$$n_1 \sin \phi_1 = \hat{n}_3 \sin \phi_3 \tag{3}$$

$\hat{r}_p$ and $\hat{r}_s$ are generally complex numbers. This is so because the electrode substrate is absorbing in the wavelength domain in question, and thus $\hat{n}_3$ is a complex number, given by $\hat{n}_3 = n_3 - ik_3$, where n_3 is the real and k_3 is the imaginary component of the complex refractive index of the metal. k is also called the "extinction coefficient." The mathematical presentation of the reflection coefficients $\hat{r}_p$ and $\hat{r}_s$ as complex numbers physically means that the electric field of the incident beam is both attenuated and phase shifted by the reflection from the electrode substrate. As seen from Eq. (2), the attenuation of the amplitude and the phase shift are different in magnitude for the p and s polarizations (other than when $\phi_1 = \phi_3 = 0°$, i.e., at normal incidence).

From Eq. (2) it can be seen that at a given wavelength, λ, the values of $\hat{r}_p$ and $\hat{r}_s$ for a given electrode material and electrolyte solution are fully defined by the optical properties of the two semi-infinite phases and by the angle of incidence, ϕ_1, i.e.,

$$\hat{r}_p, \hat{r}_s = \hat{r}_p, \hat{r}_s[n_1(\lambda), \hat{n}_3(\lambda), \phi_1] \tag{4}$$

For the case of the filmed electrode shown in Fig. 1(b), the Fresnel reflection coefficients are obtained by summing all the reflected rays originating after different numbers of reflections from the film-metal interface. The result is the Drude equation, which contains the reflection coefficients for the electrolyte-film (designated 12) and for the film-metal (designated 23) interfaces, as well as the phase delay δ due to the optical thickness of the film, [4,6]:

$$\hat{r}_p^{123} = \frac{\hat{r}_p^{12} + \hat{r}_p^{23} \exp(-i\delta)}{1 + \hat{r}_p^{12}\hat{r}_p^{23} \exp(-i\delta)} \tag{5a}$$

$$\hat{r}_s^{123} = \frac{\hat{r}_s^{12} + \hat{r}_s^{23} \exp(-i\delta)}{1 + \hat{r}_s^{12}\hat{r}_s^{23} \exp(-i\delta)} \tag{5b}$$

where $\delta = (4\pi/\lambda) d\hat{n}_2 \cos \phi_2$ and d, $\hat{n}_2$, and ϕ_2 are the thickness of the film, the complex refractive index of the film, and the angle of propagation within the film, respectively. ϕ_2 is related to ϕ_1 by Snell's law and is complex for absorbing films. The dependence of the reflection coefficients on the parameters of the system, in the case of a filmed substrate, can be described as

$$\hat{r}_p, \ \hat{r}_s = \hat{r}_p, \ \hat{r}_s(n_1(\lambda), \hat{n}_2(\lambda), \hat{n}_3(\lambda), \Phi_1, \lambda, d) \qquad (6)$$

The optical properties of an isotropic phase are described in all the equations above in terms of the complex refractive index, the imaginary component of which (k) determines the extent of light absorption at the relevant λ. Chemists are usually more familiar with a description of light energy loss in terms of the absorption coefficient, α, expressed in cm^{-1}. The relationship between the absorption coefficient and the extinction coefficient is

$$\alpha = \frac{4\pi k}{\lambda} \qquad (7)$$

Physicists usually prefer to use the complex dielectric constant, $\hat{\varepsilon}$, rather than the complex refractive index, n. The relation between the two parameters is

$$\hat{\varepsilon} = \hat{n}^2 \qquad (8)$$

and therefore,

$$\varepsilon' = n^2 - k^2 \qquad \varepsilon'' = 2nk \qquad (9)$$

where ε' and ε'' are the real and imaginary components of $\hat{\varepsilon}$. Presentations of optical equations or of results in terms of $\hat{n}$ or of $\hat{\varepsilon}$ are fully interchangeable. Most of the equations and the results will be described in this chapter in terms of $\hat{n}$, but some of the discussion of optical properties of solids will be in terms of $\hat{\varepsilon}$.

B. Ellipsometric Parameters and Their Relation to the Optical Properties of the Interface

An ellipsometer measures the ratio between the Fresnel reflection coefficients at p and s polarizations for a specularly reflecting interface. The ellipsometric parameters are designated as Ψ and Δ and are defined by

$$\tan \Psi \exp(i\Delta) = \frac{\hat{r}_p}{\hat{r}_s} \qquad (10)$$

For a two-phase substrate-ambient system [Fig. 1(a)] the ellipsometric parameters are thus given by [see Eq. (2)]

$$\tan \Psi_0 \exp(-i\Delta_0) = \frac{\hat{r}_p^{13}}{\hat{r}_s^{13}}$$

$$= \frac{(n_1 \cos\phi_3 - \hat{n}_3 \cos\phi_1)(n_1 \cos\phi_1 + \hat{n}_3 \cos\phi_3)}{(n_1 \cos\phi_3 + \hat{n}_3 \cos\phi_1)(\hat{n}_3 \cos\phi_3 - n_1 \cos\phi_1)} \tag{11}$$

and for a three-phase substrate-film-ambient system [Fig. 1(b)], the ellipsometric parameters are given by [see Eq. (5)]

$$\tan \Psi \exp(i\Delta) = \frac{\hat{r}_p^{123}}{\hat{r}_s^{123}}$$

$$= \frac{[\hat{r}_p^{12} + \hat{r}_p^{23} \exp(-i\delta)][1 + \hat{r}_s^{12}\hat{r}_s^{23} \exp(-i\delta)]}{[\hat{r}_s^{12} + \hat{r}_s^{23} \exp(-i\delta)][1 + \hat{r}_p^{12}\hat{r}_p^{23} \exp(-i\delta)]} \tag{12}$$

In the case of a two-phase electrode-electrolyte system, the dependence of the ellipsometric parameters on the optical characteristics of the system has the general form

$$\Psi_0, \Delta_0 = \Psi_0, \Delta_0[n_1(\lambda), \hat{n}_3(\lambda), \phi_1] \tag{13}$$

while for the three-phase electrode-surface film-electrolyte system, this dependence can be described as

$$\Psi, \Delta = \Psi, \Delta[n_1(\lambda), \hat{n}_2(\lambda), \hat{n}_3(\lambda), \phi_1, \lambda, d] \tag{14}$$

Frequently, the shifts $\delta\Psi$ and $\delta\Delta$ due to the surface process (e.g., film growth), defined as $\delta\Delta = \Delta - \Delta_0$, $\delta\Psi = \Psi - \Psi_0$, are used in the analysis of ellipsometric results. Obviously, $\delta\Delta$ and $\delta\Psi$ depend on the same optical characteristics of the system,

$$\delta\Psi, \delta\Delta = \delta\Psi, \delta\Delta[n_1(\lambda), \hat{n}_2(\lambda), \hat{n}_3(\lambda), \phi_1, \lambda, d] \tag{15}$$

Equations (11) and (12) have rather complex forms and do not allow an immediate intuitive "feel" for the way in which Ψ and Δ depend on the optical properties of the interface, and particularly, on the thickness and properties of a surface film. Furthermore, Eqs. (11) and (12) cannot be solved analytically. However, in this computer age this is not a significant barrier. One may approach the understanding and application of ellipsometry through the general parametic dependencies described by expressions (13) and (14), knowing that relatively simple and quite readily available computer programs exist that solve Ψ and Δ for a film-free [Eq. (11)] or for a filmed surface [Eq. (12)], given the relevant optical properties of the interface as input [8]. Furthermore, based on computer-evaluated solutions, we hope to show in Sec. IV.B that some "feel" can be developed for the way in which Δ and Ψ vary during the growth of some characteristic surface films on metallic substrates.

Let us next consider the common experimental routine that involves a measurement of Ψ_0 and Δ_0 for the film-free substrate followed by a measurement of Ψ and Δ for the same substrate when covered by the surface film investigated. Because $n_1(\lambda)$, the refractive index of the electrolyte solution, is usually known or can easily be determined, the ellipsometric measurement of Ψ_0 and Δ_0 reveals through Eq. (11) the real and imaginary components of $\hat{n}_3(\lambda)$, i.e., the optical properties of the electrode substrate. Once $\hat{n}_3$ is solved from the ellipsometric measurement of the "bare" substrate, the film should next be characterized according to the values of Ψ and Δ (or $\delta\Psi$ and $\delta\Delta$) measured for the filmed electrode, using Eq. (12). However, in the case of a nontransparent surface layer, the simplest, i.e., isotropic, film is characterized optically at each λ by three parameters: the real and imaginary components of $\hat{n}_2$ (n_2 and k_2) and the thickness of the layer, d. At the same time, the ellipsometric measurements taken at a single λ and Φ_1 supply only two measured parameters, Ψ and Δ (or $\delta\Psi$ and $\delta\Delta$), which are not sufficient to solve the three unknowns. Several possible ways to add the necessary experimental data required to solve the optical properties of the film will be described.

C.   Combined Reflectometric and Ellipsometric Measurements

One way to provide more experimental data on the filmed substrate at a given λ is to take a reflectometric measurement at the same wavelength and angle of incidence employed in the ellipsometric measurement [9]. This can be done on the same instrumental setup and with the same optical alignment of the electrode surface. It requires at most minor resetting of an optical element. The measured variable in a reflectometric measurement is the reflectance, R, defined as the ratio between the intensities of the reflected and the incident beams. For the states of polarization p and s it is given by

$$R_p = |r_p|^2 \qquad R_s = |r_s|^2 \tag{16}$$

In investigations of electrochemical systems, only the relative variations of the reflectance due to the surface process (e.g., film growth) have usually been recorded. This recording can be done by measuring the relative variations of the reflected intensity at a given state of polarization, as brought about by the surface process. Because the intensity of the reflected beam, I_{ref}, is given by

$$(I_{ref})_{p,s} = (I_{in})_{p,s} R_{p,s} \tag{17}$$

then for a constant incident beam intensity, I_{in}, the measured relative variation in reflected intensity is equal to the relative change in the interfacial reflectance, i.e.,

$$\left(\frac{\delta I_{ref}}{I_{ref}}\right)_{p,s} = \left(\frac{\delta R}{R}\right)_{p,s} = \frac{|r_{p,s}|^2 - |r_{p,s}|_0^2}{|r_{p,s}|_0^2} \tag{18}$$

where the subscript 0 refers to the film-free substrate. From Eqs. (2),
(5), and (18), it can be seen that $(\delta R/R)_{p,s}$ depends on the same
system parameters as do Ψ and Δ (or $\delta\Psi$ and $\delta\Delta$), i.e.,

$$\left(\frac{\delta R}{R}\right)_{p,s} = \left(\frac{\delta R}{R}\right)_{p,s} [n_1(\lambda),\hat{n}_2(\lambda),\hat{n}_3(\lambda),\phi_1,\lambda,d] \tag{19}$$

but the functional dependence of $(\delta R/R)_{p,s}$ on these parameters,
through Eqs. (2), (5), and (18), is different than that of Ψ and Δ
[Eq. (12)]. This makes the measurements of $(\delta R/R)_{p,s}$ complementary
to ellipsometric measurements. Expressions (14) [or (15)] and (19)
suggest that once $\hat{n}_3$ has been determined from a mesurement of Δ_0 and
Ψ_0, the four independent optical readings obtained during film growth
by combining ellipsometry and reflectometry at a given wavelength and
angle of incidence, i.e., $\delta\Psi$, $\delta\Delta$, $(\delta R/R)_p$, and $(\delta R/R)_s$, should allow
the solution of the three optical parameters of the surface film: n, k,
and d. This experimental approach has proved helpful in the analysis
of several systems, as described in detail in Sec. V.

It should be noted that reflectance measurements are different from
elipsometric measurements in that they are more sensitive to fluctuations
and drift in the intensity of the incident beam. Ellipsometric signals do
not depend, in principle, on variations in the overall intensity, because
they are obtained from ratios of reflected intensity components or are
evaluated under null conditions (see Sec. III). Thus, neither fluctuations of the incident intensity nor loss of energy due to the formation
of an absorbing product in solution affect the measurements of Ψ and Δ.
This is not the case for reflectance measurements. The assumption in
Eq. (17) that variations in I_{ref} genuinely reflect the variations of R
fails if I_{in} drifts during the measurement of the surface process, or if
attenuation of the reflected intensity is brought about by products in
the solution phase. The first problem can be solved with a split-beam
arrangement, which allows continuous monitoring and correction for
fluctuations in source intensity. The second problem limits the use of

reflectometric measurements to electrochemical processes that do not cause significant variations of light absorption in solution.

D. Analysis

Ellipsometric measurements generate Δ and Ψ readings for the film-free and for the filmed substrate. Computer programs for the solution of Ψ, Δ, and $(\delta R/R)_{p,s}$ for given physical parameters and experimental conditions according to Eqs. (11), (12), and (19), respectively, have been written and are available [8]. The computer programs typically solve first the optical properties of the substrate n_3 and k_3, from the readings Ψ_0, Δ_0 for the film-free state. These are then used to generate solutions of Δ, Ψ (or $\delta\Delta$, $\delta\Psi$) and $(\delta R/R)_{p,s}$ for any chosen values or range of values of film properties, n_2, k_2, and d. However, to solve an unknown surface film, a solution of the reverse problem is required, i.e., evaluation of n_2, k_2, and d according to the measured changes in ellipsometric parameters. Some of this has been done previously by visual comparison of the experimental results for film growth, usually presented in the form of a Ψ vs. Δ plot, to a family of Ψ vs. Δ plots calculated for the growth of films with a range of optical properties. In arriving at such solutions, some knowledge of the typical response of the Δ and Ψ parameters to the growth of films with certain characteristic properties helps in estimating the approximate film properties (see Sec. IV.B).

A more reliable approach, which has been adopted more recently (see, e.g., Refs. 10 to 12 and 106), is based on a computer search for the best fit of the experimental results to the simulated optical effect of a specific film. If only ellipsometric readings are taken at several stages in the uniform growth of an investigated film, such a computer search can be performed by minimizing the function

$$F = \sum_{i}^{N} \left(\Delta_{calc,i} - \Delta_{expt,i} \right)^2 + \sum_{i}^{N} \left(\Psi_{calc,i} - \Psi_{expt,i} \right)^2 \qquad (20)$$

where $\Delta_{expt,i}$ and $\Psi_{expt,i}$ have been measured at stage number i of
film growth. Such minimization yields the real and imaginary components
of the refractive index for the film in question n_f, k_f, and the N film
thicknesses. In such a case, this information is obtained with no a
priori assumption about either the optical properties or about the film
thicknesses, other than the assumption of uniform film growth. This
last assumption, i.e., that the optical properties of the film do not
depend on film thickness in the range investigated, can be verified (or
discredited) by the quality of the best fit between the experimental
results and a computer-generated plot based on the assumption of film
uniformity. It can be seen that for N stages of growth, the number of
unknowns for a uniform isotropic layer is $2 + N$ (n_f, k_f, and N thicknesses) and the number of measured parameters is 2N (a Ψ, Δ pair for
each stage). Thus the data become sufficient for $N > 1$. An example
for such a fit is shown in Fig. 2.

In the case of a film that can be grown only to a single unique
thickness or of a film with optical properties that vary significantly
with thickness, another approach to the analysis is required. The use
of a function like Eq. (20), but with measured and calculated values of
$(\delta R/R)_{p,s}$ also included in the minimization, allows us in principle to
find the best solution for n_f, k_f, and d, according to optical readings
obtained at a single film thickness (see Sec. II.C). A similar least-
squares analysis can be used also to solve the properties of a film,
according to ellipsometric reading taken for a given λ at several angles
of incidence [13].

Use of additional assumptions about the nature of the surface film
has been common in the analysis of ellipsometric results. Typical
assumptions in electrochemical work have been that the thickness can be
estimated from coulometric data or that the refractive index of the film
can be assumed to be identical to that of a known bulk compound.
Both assumptions can lead to substantial errors. The significant

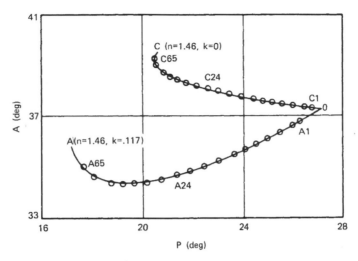

FIG. 2. Computer-aided fit of experimental ellipsometric data to a model
that assumes uniform film growth. Points C_i correspond to the cathodic
form and points A_i to the anodic form of an iridium oxide film, grown
in stages by oxidation-reduction cycles. The best fit corresponds to
$\hat{n}_f = 1.46-0i$ for the cathodic form and $\hat{n}_f = 1.46-0.117i$ for the anodic
form, at 633 nm. The metal substrate properties evaluated ellipsometrically were: $\hat{n}_{Ir} = 2.512-5.085i$. To convert the ellipsometric null readings P and A, which appear in the figure to Δ and ψ values, the simple
relations are in this case: $\Delta = 90 + 2P$; $\psi = A$, as explained in Sec.
III.B. (From Ref. 51. Reprinted by permission of the publisher, The
Electrochemical Society, Inc.)

possible water content in surface films formed in electrochemical systems
is one reason that makes the connection between the charge passed during film growth and the thickness of the film ambiguous, even in the
rare cases in which the faradaic efficiency for growth is very close to
100%. A priori assumptions about the optical properties are always
questionable because of the unknown water content as well as of the
unknown degrees of nonstoichiometry and of structural and compositional
imperfections. It becomes clear that the approach to analysis of the
optical data should be complete reliance on the optical results. For ·thin
films (<10 nm), particularly when only a single film thickness can be
probed, this usually means the use of reflectometric measurements to

help in the analysis. For thicker films which can be grown uniformly
to different thicknesses, a complete analysis can be based on ellipsometric results alone without additional assumptions [Eq. (20)]. In
both cases, the computer programs that search for the best fit between
mesaured and calculated parameters have proved to be most efficient.

Finally, it should be made clear that in many cases a single film is
not a sufficient model for the interfacial process (see Sec. V). In
cases where two films, or more, are required for a full optical description of interfacial processes, the system naturally involves more parameters and requires more measurements, e.g., at a larger number of film(s)
growth stages, for complete characterization.

E.   Spectroscopic Ellipsometry

In any measurement of the response of an investigated system to a
perturbation, the response function in a range of frequencies is much
more informative than the response of the system at a single frequency.
In the same way, the dispersion of the complex refractive index (or
complex dielectric constant) for an investigated surface layer contains
much more information than does the single value of the complex refractive index at a single λ. The most recent applications of ellipsometry
in electrochemistry show a clear transition from single λ ellipsometry to
automatic spectroscopic ellipsometry (see Sec. V). No new principles
are involved in the extension of the technique to cover a range of wavelengths. As λ is scanned in a spectro-ellipsometer, readings of Ψ and Δ
are collected as a function of λ, and yield n_2, k_2 and d as a function of
λ according to analysis based on Eqs. (11) and (12) and on the approaches described in Sec. II.D. The analysis is performed separately
for each λ, e.g., in a range of evenly spaced discrete values of λ,
according to the ellipsometric readings collected. One advantage of a
large number of data points collected in a wide λ range is that the
thickness of the film, d, can be assumed constant in the analysis as

long as the interfacial process can be described by the growth of a single film. On the other hand, if independent evaluation of film thickness at each λ yields significant variations of d with λ, this suggests that the description of the optical behavior in terms of a single uniform film is not satisfactory. But the most important advantage of ellipsometric measurements taken in a wide λ range is obviously the resulting spectrum (or "dispersion") of the optical properties in the near IR-visible-near UV region, which allows much better insight into the nature of the electronic transitions contributing to the optical behavior of the film, as well as the implications with regard to the conductivity of the film under dc conditions (see Sec. IV.A). Such spectra are sometimes very revealing also with respect to the microstructure of the film (see Sec. V.C). For instrumentation, spectroscopic ellipsometry requires either a completely "achromatic" measuring system, in which the optical function of each of the elements does not depend at all on λ, or an automatic adjustment of the setting of one of the elements during the λ scan. This is possible with any of the common versions of automatic ellipsometers, as described in Sec. III.

III. INSTRUMENTATION

A. How Ψ and Δ Are Measured: The Rotating-Analyzer Ellipsometer

Instrumentation in ellipsometry has very clearly changed since 1973 [4,5]. Practically all of the work done in recent years on electrochemical systems has been performed with automatic ellipsometers, as opposed to the classical and slow null ellipsometers employed earlier by most workers in this field. Automatic ellipsometers allow time-resolved measurements with millisecond, or even submillisecond resolution and fast collection of data for a wide wavelength range. Automatic instruments are available commercially and are usually provided without λ scanning capabilities. However, adding such a capability is not a formidable task. In several cases the advent of minicomputers and microprocessors allowed research groups to build their own automatic ellipsometric systems, thus allowing more flexibility and achieving some special desired features.

The measurement of the ellipsometric parameters Δ and Ψ for a reflecting surface can be explained in a relatively straightforward way through the description of the rotating-analyzer version of automatic ellipsometry [14—17]. This version has been employed quite frequently in research in interfacial electrochemistry [17—19]. (Some commercially available automatic ellipsometers designed for research work, e.g., those supplied by Rudolph Research or by Gaernter Scientific Corp., employ a rotating-analyzer.) Figure 3 shows a common configuration of the rotating-analyzer (RA) ellipsometric system. The incident collimated monochromatic beam is linearly polarized, with the plane of polarization usually at 45° to the plane of incidence. For this specific state of

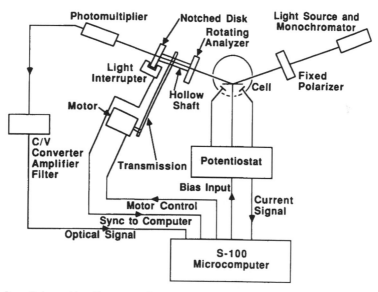

FIG. 3. Schematic diagram of an automatic rotating-analyzer ellipsometer (see Ref. 17).

polarization, the ratio between the electric field components in the p and s directions in the incident beam, is exactly 1, and the phase difference is exactly zero, i.e.,

$$\left(\frac{E_p}{E_s}\right)_{in} = 1 \tag{21}$$

The ratio of the complex electric field vectors along the p and s directions in the reflected beam will be determined by the ratio of the Fresnel reflection coefficients and hence will generally depend on the ellipsometric parameters of the interface according to [see Eqs. (1) and (10)]

$$\left(\frac{\hat{E}_p}{\hat{E}_s}\right)_{ref} = \left(\frac{E_p}{E_s}\right)_{in} \frac{\hat{r}_p}{\hat{r}_s} = \left(\frac{E_p}{E_s}\right)_{in} \tan \Psi \exp(-i\Delta) \tag{22}$$

and when $(E_p/E_s)_{in} = 1$,

$$\left(\frac{\hat{E}_p}{\hat{E}_s}\right)_{ref} = \tan \Psi \exp(-i\Delta) \tag{23}$$

Thus, for this special case of an incident beam linearly polarized at 45° to the plane of incidence, analysis of the reflected beam to give the complex ratio $(\hat{E}_p/\hat{E}_s)_{ref}$ would directly yield the ellipsometric parameters Ψ and Δ for the reflecting interface. What does such an analysis mean? Due to the phase difference between E_p and E_s which is introduced by the reflection, the reflected beam is generally elliptically polarized; i.e., the tip of the electric field vector rotates in an elliptical orbit at a frequency of the order of 10^{15} Hz in a plane perpendicular to the propagation of the reflected beam. Such high frequency of rotation means that the ellipse described by the tip of the E_{ref} vector can be treated as if it were stationary. Analysis of the complex ratio $(\hat{E}_p/\hat{E}_s)_{ref}$ is equivalent to the characterization of this ellipse. In the rotating-analyzer mode of ellipsometry, the characterization of the elliptically polarized reflected beam is performed by passing it through a liner polarizer ("analyzer"), which rotates at a constant angular velocity (typically of the order of 100 rpm) in the plane perpendicular to the beam, thus sampling the component of the reflected intensity along the orientation, θ, of the analyzer's optical axis. The intensity of the light emerging from the rotating analyzer will exhibit a sinusoidal dependence on 2θ, and because θ varies linearly with time, the intensity will vary sinusoidally with time as well. The time corresponding to the orientation $0°$ (p-vector) of the rotating analyzer is probed, e.g., by the notched disk and the light interrupter, as shown in Fig. 3. Figure 4(a) shows the Ellipse described by the E_{ref} vector as seen looking into the reflected beam, and Fig. 4(b) shows the characteristic parameters of the sinusoidal intensity waveform resulting from the passage of the elliptically polarized reflected beam through the rotating analyzer. The ratio of minimum to maximum ("valley to peak") in this sinusoidal intensity waveform is designated by tan γ and is equal to the square of the ratio between the minor and major axes of the E_{ref} ellipse. The phase of the intensity waveform, designated by α, will be determined by the orientation of the major axis of the ellipse with respect to the plane of incidence (or the p-vector). Tan γ and α define the ellipse in its "natural" cartesian system determined by its axes. What is left to do to derive Δ and Ψ is to translate the description of the ellipse from this "natural" system to the "physical" cartesian system defined by the plane of incidence, i.e., by the p and s vectors. From simple coordinate transformation considerations it can be shown that [4,6]

$$\tan \Delta = -\frac{\tan 2\gamma}{\sin 2\alpha} \tag{24a}$$

$$\cos 2\Psi = \cos 2\gamma \cos 2\alpha \tag{24b}$$

where

$$\tan \gamma = \frac{I_{min}}{I_{max}} \tag{25a}$$

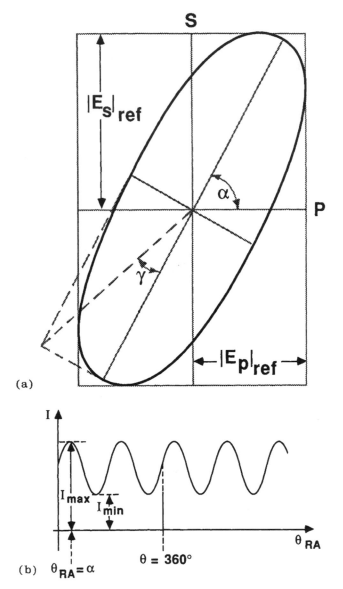

(a)

(b)

FIG. 4. (a) Ellipse described by the tip of the E_{ref} vector in an elliptically polarized light beam, shown with the characteristic parameters α and tan γ, which define it in the cartesian system of the ellipse axes. Also shown are the amplitudes of the components of E_{ref} along the p and s orientations, the combination of which with a phase difference Δ results in the same elliptical rotation. $|E_p| / |E_s|$ = tan Ψ. (b) Light intensity vs. analyzer orientation for the elliptically polarized reflected beam, as detected after passing a rotating analyzer.

and

$$\alpha = \theta_{I_{max}} \tag{25b}$$

as explained for the rotating-analyzer measurement (see Fig. 4). Thus, converting the sinusoidal intensity variation in the beam emerging from the rotating analyzer to a sinusoidal voltage signal with the aid of a linear photodetector, and evaluating $\tan \gamma$ and α from this sinusoidal voltage waveform with appropriate electronic circuitry, we can derive Ψ and Δ.

Some of the rotating-analyzer automatic ellipsometers [14] have indeed directly measured $\tan \gamma$ and α. However, with the recent development of microprocessors, evaluation of the ellipsometric parameters from the sinusoidal intensity waveform with a digital phase-sensitive detector has become a preferred routine. Digital processing of the sinusoidal intensity waveform emerging from the rotating analyzer can be performed by either Fourier [15,16] or Walsh [17] transform techniques. The revelant presentation of the intensity wave form is then in terms of three Fourier coefficients, where the intensity I_{ref} is given by [16,17]

$$I_{ref} = c_0 + c_1 \cos 2\theta + c_2 \sin 2\theta \tag{26}$$

where θ is the orientation of the rotating analyzer with respect to the plane of incidence. Evaluation of c_0, c_1, and c_2 is performed by a microprocessor that takes readings of the instantaneous intensity at 2^N points (N, an integer) equally spaced in time during each period of the sinusoidal signal. This microprocessor digitizes these readings and performs a Walsh transform (or a fast Fourier transform). In the simplest case, in which Eq. (21) applies, Ψ and Δ can be calculated as [16]

$$\tan \Psi = \frac{c_0 + c_1}{c_0 - c_1} \tag{27a}$$

$$\cos \Delta = \frac{c_2}{(1 - c_1^2)^{1/2}} \tag{27b}$$

As pointed out in Sec. II.D, reflectometric measurements provide
further data that help in the analysis of the surface film examined. In
automatic ellipsometers based on a rotating analyzer, variations of the
reflectance can be directly recorded on the same setup simultaneously
with the recording of Δ and Ψ. From Eq. (26), it can be seen that c_0
is proportional to the reflectance of the surface for the incident light
employed. For light linearly polarized at 45°,

$$c_0 = \frac{1}{2} (R_p + R_s) I_{in} \tag{28}$$

and since c_0 is evaluated by the digital transform of the intensity waveform, the relative variations in the sum $R_p + R_s$ are readily derived
with a rotating-analyzer automatic ellipsometer. Individual changes in
R_p or in R_s can be measured on the same setup in separate experiments
if the incident light is polarized at either 0°(p) or 90°(s) and the
intensity of the reflected beam is followed during the surface process
monitored. As noted above (Sec. II.C), variations in the reflected intensity resulting from source fluctuations are automatically corrected in
the evaluation of Δ and Ψ, since both are given by ratios of Fourier
coefficients [Eqs. (27a) and (27b)]. But this is not the case for the
reflectance.

Because only linear polarizers are used in ths ellipsometric system,
the rotating-analyzer version of automatic ellipsometry allows scanning
of the wavelength and collection of Δ and Ψ spectra without the need for
adjustment of any of the elements as the wavelength of the incident
beam is scanned.

In a recent contribution, Rishpon et al. described the use of a low-
cost microcomputer for the automation of ellipsometry in the rotating-
analyzer mode [17]. Using a simple dc motor and a transmission belt,
stabilization of the motor driving the rotating analyzer, by means of a
computer-controlled on-off switching mechanism, allowed stabilization of
the rotational velocity (250 rps) to within 1 part in 10^4. This allowed,

in turn, measurements with time resolution of 0.01 sec. A precision of 0.01° for a pair of Ψ and Δ readings following 1 sec of integration time was obtained by employing a simple and fast Walsh transform in the digital analysis of the sniusoidal intensity waveform. For further details on instrumentation of RA ellipsometers, including precision and error analysis, the reader is referred to Refs. 14 to 17, 20, and 21.

B. Classical Ellipsometric Null Measurement and Self-Nulling Automatic Ellipsometers

The classical approach to the measurement of Δ and Ψ has been by means of a null technique (see, e.g., Refs. 4 and 6). The principle of this technique is to introduce ellipticity in the incident beam, such that following reflection from the surface examined, the difference in phase shifts for the p and s polarizations will lead to a linearly polarized reflected beam that can be perfectly extinguished with a single linear polarizer. In its most common version, the measurement involved passing the incident beam through a linear polarizer and next through a "compensator" made of a birefringent material. The last element introduces a phase difference (usually $\pi/2$) between the components of the electric field vector along the mutually perpendicular "fast" and "slow" axes. The beam is next incident on the sample and after the reflection, passes through another linear polarizer ("analyzer") and reaches a linear photodetector. The orientations of the optical axes of all the elements (P for the polarizer, C for the compensator, and A for the analyzer), measured with respect to the plane of incidence, are manually adjustable. For the case of a constant azimuth of 45° of the compensator, the measurement involves finding the angular readings of the polarizer and analyzer, P_{null} and A_{null}, for which the intensity of light reaching the photodetector is extinguished ("null point"). Under such conditions, it can be shown that Ψ and Δ for the reflecting surface are related to one pair of the null readings P_{null} and A_{null} according to

$$\Delta \fallingdotseq 270° - 2P_{null} \qquad (29a)$$

and

$$\Psi = A_{null} \qquad (29b)$$

As shown by the example in Fig. 2, additional different pairs of null readings are obtained in other quandrants [4,6], but they should obviously all yield the same pair of Δ and Ψ for the reflecting surface. The manual search for a minimum of the intensity is a lengthy procedure, of the order of a minute per point. However, the classical null technique is very accurate and can thus serve for occasional checks on the results obtained with automated instruments. For further details on the classical null approach, the reader is referred to Refs. 1, 4, and 8.

Automatic self-nulling ellipsometers [22,23] operate on the same principle as that of a classical null ellipsometer. A detailed description of such a self-compensating ellipsometer, with the capability of spectral scanning, has recently been given by Muller and Farmer [23] in the context of ellipsometric work on electrochemical systems. Its scheme is given in Fig. 5. Automated nulling is achieved in the following way: The azimuth of the linearly polarized incident beam that enters the compensator is determined by the sum of a manually adjusted angular reading in a regular polarizer and an additional electronically controlled rotation induced by a Faraday cell. Another Faraday cell is introduced between the sample and the analyzer in the reflected beam. The overall analyzer and polarizer null readings A_{null} and P_{null}, are thus given by

$$A_{null} = A_{Far} + A_{man} \qquad (30a)$$

and

$$P_{null} = P_{Far} + P_{man} \qquad (30b)$$

where A_{man} and P_{man} are the angular readings on the manually adjusted units, and A_{Far} and P_{Far} are the readings of the electronically controlled

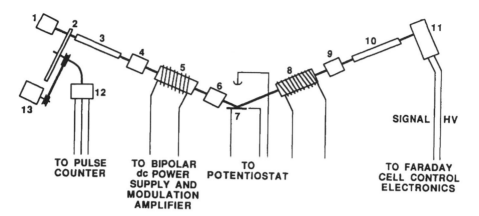

FIG. 5. Spectral-scanning, self-nulling ellipsometer, optical components, and electrical connections: (1) high-pressure Xe lamp; (2) rotating interference filter; (3) collimator; (4) polaizer Glan-Thompson prism; (5) polarizer Faraday cell' (6) Fresnel rhomb achromatic retarder; (7) electrochemical cell; (8) analyzer Faraday cell; (9) analyzer Glan-Thompson prism; (10) collimator; (11) spectrally "flat" detector; (12) rotary incremental digital encoder; (13) spectral scanner drive motor. (From Ref. 23.)

Faraday cells. Ψ and Δ will depend on A_{null} and P_{null}, respectively, according to Eq. (29), in the same way as in a classical (null) ellipso-meter. The obvious advantage over the classical ellipsometer is that the instrument has now become self-compensating: It can automatically search for the values of A_{Far} and P_{Far} which bring about minimum intensity as detected by the photodetector. To achieve wavelength scanning capa-bility, an achromatic compensator, i.e., a compensator that does not require continuous adjustment during a λ scan, is needed. Muller and Farmer used a Fresnel rhomb for this purpose (Fig. 5).

   C.   Automatic Ellipsometers Based on the Periodic Modulation
          of the Polarization in the Incident Beam

Two instruments of this last type have recently been described in the electrochemical literature [24,25]. The basis of this instrumental ap-proach is sinusoidal modulation of the ellipticity of the incident beam, generated by a standing strain wave of high frequency, e.g., 50 kHz, in a fused silica block. The strain wave is induced by a piezoelectric transducer that is cemented to one end of the silica block and that is powered by a 50-kHz electrical signal. The name for this commercially available unit is "photoelastic modulator." In the complete optical setup shown in Fig. 6, the incident beam passes first through a linear polarizer and then through the photoelastic modulator before reaching the sample surface examined. The reflected beam passes through one linear polarizer ("analyzer") and is probed by a linear photodetector. For photoelastic modulation at frequency ω (e.g., 50 kHz), the reflected beam contains intensity components at ω and 2ω (50 and 100 kHz) and a dc intensity component. It can be shown [26] that for the settings

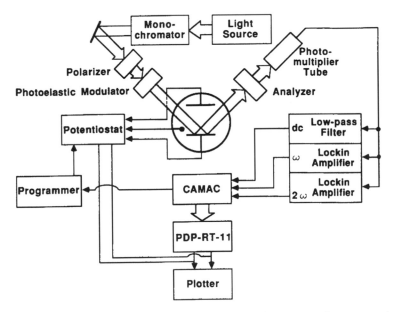

FIG. 6. Schematic diagram of an automatic ellipsometer for measuring electrochemical interfaces, based on polarization modulation at 50 kHz by a photoelastic modulator. (From Ref. 25. Reprinted with permission of Elsevier Science Publishers, Amsterdam.)

$P = 90°$, $C = 45°$, $A = -45°$, the ratios of these frequency components in the reflected beam depend on Δ and ψ of the reflecting surface, as in

$$\frac{(I_\omega/I_{dc})_{ref}}{R'_{cal}} = \sin 2\Psi \sin \Delta \qquad (31a)$$

and

$$\frac{(I_{2\omega}/I_{dc})_{ref}}{R''_{cal}} = \cos 2\Psi \qquad (31b)$$

R'_{cal} and R''_{cal} are designations for the ratios $(I_\omega/I_{dc})_{cal}$ and $(I_{2\omega}/I_{dc})_{cal}$, respectively, recorded under properly defined calibration conditions [26]. R'_{cal} and R''_{cal} in Eq. (31) depend on the amplitude of the photoelastic modulation as well as on amplifier gains and other instrument-dependent constants, but they do not depend on the properties of the reflecting sample. The calibration is typically performed once at the beginning of a day of experiments. To read the ratios $(I_\omega/I_{dc})_{ref}$ and $(I_{2\omega}/I_{dc})_{ref}$, the signal from the photodetector is fed to two lock-in amplifiers and to a dc amplifier. The ω, 2ω, and dc components can thus be followed as a function of time (or potential) during the evolution of a surface process or during a wavelength scan. The calculation of the ratios between the components of the intensity and the evaluation of Ψ and Δ for the surface at each point on a time, potential, or wavelength axis can be made according to Eqs. (31b) and (31a) following the completion of a single scan or of multiple scans. This is most effectively performed with the aid of an on-line computer that digitizes the readings of I_ω, $I_{2\omega}$, and I_{dc} and performs the calculations of the ratios in Eq. (31) and of Ψ and Δ as a function of the variable described. The on-line computer also allows averaged redings in repeated scans (of potential or λ) and thus is able to enhance the signal/noise ratio [25]. During a linear λ scan, the amplitude of the photoelastic modulation has to be linearly scanned to adjust it for the varying wavelength. This is easily done with a linear voltage ramp input to the power supply of the modulator. Relative variations in the reflectance can also be recorded directly with this instrumental setup by following the dc components of the reflected intensity I_{dc} (see Eq. 4 in Ref. 26). For more information on automatic ellipsometers based on polarization modulation, see Refs. 26 to 28.

D. Accuracy, Precision, Response Time, and Spatial Resolution

Candela and Chandler-Horowitz [29] have made the most elaborate recent effort to achieve high accuracy in ellipsometric measurements. They built a system for accurate metrology of thin films. The system comprises a computer-controlled spectroscopic ellipsometer, to be used for the metrology of semiconductor materials and the calibration of reference standards for thin-film thickness and refractive index. Because maximum accuracy was required, a substantial effort was invested in the mechanical part of the system and in the automatic alignment procedure. The system employs both a conventional null ellipsometer and a rotating-analyzer ellipsometer on the same optical bench. The authors report a standard deviation of 10^{-4} degree for Δ, Ψ readings by the classical null method, and a similar precision achieved with a special version of the rotating-analyzer technique (principal-angle RAE). In most ellipsometric-electrochemical investigations, the accuracy obtained is less impressive, in large part because of a significantly larger uncertainty in the angle of incidence, as compared with the 10^{-3} degree achieved in the system mentioned above [29]. Another important reason for inaccuracies in Δ, Ψ redings is the irreproducibility in the surface preparation of metal electrode surfaces, those mechanically polished, evaporated, or sputtered. However, the requirements for accuracy in electrochemical systems are usually not as stringent. Depending on the system investigated, a precision of the order of $10^{-3}-10^{-2}$ degree is usually sufficient. The stress here on precision rather than accuracy results from the focus on investigations of surface films grown on electrode substrates. The ellipsometric effects due to film growth are difference parameters, i.e., $\delta\Psi$ and $\delta\Delta$, which are not strongly affected by small inaccuracies in the determination of Ψ_0, Δ_0, for the "bare" substrate. For example, accuracy in Δ_0 and Ψ_0 of ±0.5 degree was shown [30] to bring about an error of less than 1% in the values of n and k for an oxide film on a metal substrate, assuming precisely determined ($\pm 10^{-2}$ degree) $\delta\Psi$ and $\delta\Delta$ due to film growth. Comparable precisions have been achieved with different automated versions of ellipsometry. For example, Rishpon et al. [17], have reported a precision of 10^{-2} degree following a 1-sec integration time with a relatively simple rotating analyzer instrument. Hyde et al. [25] showed that precision of the order of 10^{-3} degree can be achieved with a polarization-modulated ellipsometer in measurements of effects with a total amplitude of 10^{-2} degree within the double-layer region, following averaging for 15 min. The performances of the two systems are thus quite comparable. A significantly better reproducibility can be obtained in ellipsometric measurements of electrochemical systems in which the surface process is controlled by continuous modulation of the electrode potential. In such cases, repetitive triangular potential scanning keeps the surface clean of impurities and allows collection of multiple signals of Δ, Ψ vs. V for improvement of the signal/noise ratio [25]. For further details on the estimation of accuracy and precision of various ellipsometric systems, the reader is referred to Refs. 1, 22, 23, 28, 29, 31, and 32. Reference 31 deals specifically with uncertainties in ellipsometric film-thickness determinations.

Most automatic ellipsometers are associated with a response time of 1—10 msec [17,20,23,25]. Moritani et al. [33,34], have reported on an automatic ellipsometer based on an ADP electro-optic modulator, with a modulation frequency of 25 MHz, which allows a data collection rate of 4 μsec per point. These authors point out, however, that "meaningful" time intervals per point have to be at least 2msec long to bring the signal/noise ratio to a reasonable level [33]. On the other hand, it can

be argued that in a repetitive triangular potential modulation experiment
a time resolution of, e.g., 20 μsec can be maintained while the requirement of 2 msec collection time per point can be fulfilled by repeating the
scan 100 times. These recent acheivements make the use of various off-
null techniques [35—39] almost obsolete. Nevertheless, in some cases
those earlier extremely simple approaches to time-resolved ellipsometry
of very thin films generated results that turned out to be almost identical to those obtained later with automatic ellipsometers (see, e.g., Refs.
39 and 25). Recently, Jellison and Lowndes described time-resolved
ellipsometric measurements capable of time resolution well below 1 μsec
[133]. Their instrument uses the familiar polarizer-compensator-sample-
analyzer arrangement; the analyzer chosen was a Wollaston prism. This
type of prism allows two channels of reflected intensity data to be
collected simultaneously. The function $F = (I_1 - I_2)/(I_1 + I_2)$, where
I_1 and I_2 are the intensities of the two beams from the Wollaston prism
analyzer, will be equal to $\cos 2\Psi$, $\sin 2\Psi \sin \Delta$, or $\sin 2\Psi \cos \Delta$, depending on the azimuthal angles of the polarizer, compensator, and analyzer.
The authors demonstrated measurements of these three functions (in
three separate experiments) with 100 nsec time resolution, during and
after pulsed laser irradiation of a Si sample [133]. The authors claim
that picosecond time resolution is achievable with this instrumental approach. It is worthwhile noting that at least two consecutive experiments
are required to obtain both Ψ and Δ with this instrumental approach.
Thus good reproducibility of the surface process investigated in two
consecutive experiments is required to take full advantage of this
impressive breakthrough in time resolution.

Conventional ellipsometry measures an area of the examined surface
which is typically a few square millimeters. To increase the spatial
resolution, Sugimoto and Matsuda have developed a microscopic ellipsometer that allows determination of the thickness and optical properties
of a thin film on an area about 10 μm in diameter [134]. The special

characteristic of this ellipsometer is that it is equipped with two groups
of lenses behind the analyzer for magnifying the polarized reflection
image of the specimen, in addition to a common polarizer-compensator-
analyzer configuration. A maximum magnification of ×600 is possible,
allowing measurement of Δ and Ψ for regions about 10 μm in diameter.
Sugimoto et al. [135], have recently described the use of this instrument
to generate profiles of passive film thickness on stainless steel samples
in NaCl solutions. They demonstrated the capability to measure the
topography around individual pits, generating such topographical maps
with a resolution of 20—50 μm. These results are discussed further in
Sec. V.E.

IV.   OPTICAL PROPERTIES OF SURFACE FILMS IN ELECTROCHEMICAL SYSTEMS AND THEIR QUALITATIVE ELLIPSOMETRIC EFFECTS

A.   Optical Properties of Surface Films in Electrochemical Systems

We have explained in Sec. II how the thickness and complex refractive
index of a surface film can be derived from the ellipsometric and reflectometric parameters. In Sec. III we described how the ellipsometric and
reflectometric parameters can be measured automatically in situ. We
next discuss briefly the nature of the information provided by such
optical data and in what ways the information clarifies the behavior of
an electrochemical interface. Independent information on film thickness,
which does not rely on assumptions on the faradaic efficiency of film
growth and the stoichiometry of the film, is in itself very valuable.
Film thickness determination can be made ellipsometrically with an accuracy of about ±5% for films significantly thicker than 10 nm, and accu-
racies of ±10% can be obtained from ellipsometry for thinner films in the
range 2—10 nm, provided that the refractive index is known accurately
[31]. The accuracy in thickness determination may be somewhat less
impressive when $\hat{n}_f$ is not known and when the assumption of perfectly

uniform film properties is not fulfilled. For a thorough treatment of this problem, see Refs. 40 and 41.

The optical properties measured by ellipsometry are determined by the nature of the electronic states in the surface phase formed on the electrode and by the nature of the interaction of the electromagnetic probe with these states. It would seem that the most straightforward way to analyze such optical data is to obtain the widest possible spectrum of $\hat{n}_{film}$ (or $\hat{\varepsilon}_{film}$), with the hope of comparing it directly with electronic spectra of known well-defined materials. Features of electronic spectra are not always easily used as a "fingerprint" because of their large width. Nevertheless, examples such as the identification from spectroscopic ellipsometry of Pb(II) oxide as a component in the anodic film formed on lead [42] or identification of the anodic oxide film formed on iron as hydrated Fe_2O_3 [18] from the ellipsometrically evaluated spectrum have been reported. An important limitation of direct comparisons between optical properties of surface films and of known bulk materials is that electrochemically formed films are more often than not composite media with intermixed solid and electrolyte components. However, with tools for realistic analysis of the ellipsometric data in the form of the appropriate version of effective-medium theory, the optical spectrum of a composite surface layer can yield valuable information on the microstructure in the film, as discussed in detail in Sec. V.C.

Electrochemists using ellipsometry often want to know whether a surface film can be classified as an insulator, a semiconductor, or a metal. This classification has an obvious bearing on the electrochemical behavior of the interface and has thus been of interest in the applications of ellipsometry in electrochemical systems. In most cases, such classification can be attempted based on the magnitude of k_f, the extinction coefficient of the film. Insulators are usually associated with a completely filled band of electronic states (valence band) separated by a large gap (>4 eV) from the lowest unoccupied band of electronic states (the conduction band). The result is that the extinction coefficient at energies well below the interband transition, E_g, is expected to be very small. Insulators are thus usually transparent and associated with values of $k_f = 10^{-2}$ or below in the visible and near IR, corresponding to an absorption coefficient $\alpha = 10^3$ cm^{-1} or less. Typical levels of (real) n for such materials is 1.5–2.5. On the other end of the scale, materials with high electronic conductivity contain a significant population of delocalized electrons. The simplest presentation of the optical properties of an "electron gas," representing free electrons in a solid according to the classical Drude model [43], is shown schematically in Fig. 7. This model predicts that at sufficiently long wavelengths (which usually

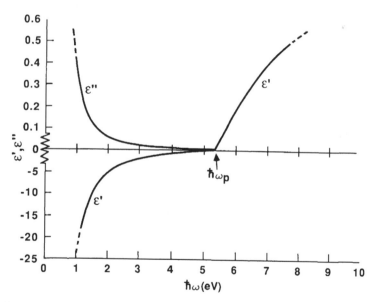

FIG. 7. Spectrum of the real (ε') and imaginary (ε'') components of the complex dielectric constant for an "electron gas," representing free electrons in a solid according to the Drude model (see, e.g., Ref. 43). The specific values of the model parameters are in this case: $\omega_p = 5.48$ eV, where ω_p is the plasma frequency; $h\tau^{-1} = 0.02$ eV, τ being the mean free time between electron collisions.

correspond to the long-wavelength end of the visible or to the near-IR region for real metals), the value of ε' is expected to be large and negative, while the value of ε'' is expected to be much smaller and positive; both components of ε should increase indefinitely in absolute value with λ. The corresponding values of k_f are typically larger than 1 ($\alpha \geq 10^6$ cm^{-1}), and since $\varepsilon' < 0$, $k_f > n_f$. Optical transitions due to free electrons of a volume density typical for metals are thus frequently identified by the large extinction coefficients ($k_f \geq 1$) in the visible. In both metals and semiconductors, interband transitions contribute strongly to the electronic spectrum. The main valence band-to-conduction band transition in a semiconductor is characterized by an "absorption edge." On the high-photon-energy side of this edge, the absorption coefficient, α, is typically 10^3-10^5 cm^{-1}, i.e., values of k_f as large as 0.1 are expected at $h\nu > E_g$. In addition to transitions between major bands, other optical transitions between narrow bands or between localized states may bring about similar levels of k_f. A salt or molecular film, where the ionic or molecular building blocks are associated with a "chromophore," may also exhibit $k_f \sim 0.1$. A model for the spectrum of n due to a transition between two electronic states was given by Lorentz [43], and this is described schematically in Fig. 8. The model is characterized by a somewhat distorted gaussian for k and a dispersion of n which resembles the derivative of the dispersion of k. It is also worthwhile noting that the real part of n dips as the peak of k is approached from the high photon energies, but increases and remains larger on the low-photon-energy side of the k peak.

To summarize this discussion, it should be noted that in real materials, the optical constants in a given wavelength region are determined by the superposition of dispersions due to one or more interband transitions (Lorentz model), plus the dispersion due to intraband transitions (Drude model) when delocalized electrons are present. It can be concluded that to validate the presence of a "metallic film" from

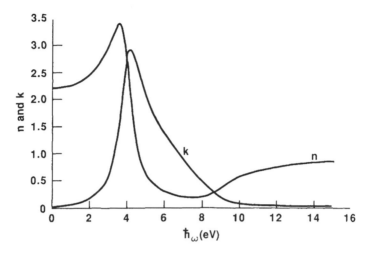

FIG. 8. The spectrum of the real (n) and imaginary (k) components of the complex refractive index for a Lorentz oscillator (see, e.g., Ref. 43). The specific values of the model parameters are, in this case: $\hbar\omega_{0'} = 4$ eV, where ω_0 is the oscillator resonance frequency; $\hbar\Gamma = 1$ eV, where Γ represents oscillator damping and provides for an energy loss mechanism; $4\pi Ne^2/m = 60$, where N is the volume density of the electronic oscillators, e the electron charge, and m the electron mass.

optical measurements in the near UV, visible, and near IR, it is important to probe the magnitudes of n_f and k_f (k_f typically ≥ 1 and $k_f > n_f$), and to verify the presence of the characteristic behavior due to free electrons, i.e., a negative ε' increasing continuously in absolute value for increasing λ. If the magnitude of k_f is smaller than the typical level for metals, but still quite large, e.g., 0.05—0.5, the absorption in the film is most probably due to interband transitions or to transitions between localized states. The nature of the spectrum of k_f can determine whether the energy loss is due primarily to a band-to-band transition, in which case the band gap for a semiconducting film can be found, or is due to a transition between localized electronic states, as would be the case for a film that contains discrete highly absorbing molecular units. It is obvious that to obtain more complete information on the electronic properties of the film, spectroscopic ellipsometry has to be employed, to reveal not only the magnitude, but also the spectral features, of $\hat{n}_f$.

Finally, it should be remembered that because ellipsometry measures not only the energy loss (absorption) but also the real part of the complex refractive index, valuable information on the film can be obtained from ellipsometric measurements of films that are totally nonabsorbing in the wavelength region employed. The value of (real) n for such films is a measure of film density and water content, and this value can be used as a good criterion to distinguish between dense oxide films, which serve as good barrier layers (high n_f), and porous hydrous, oxyhydroxide, or hydroxide structures (low n_f). Such examples are given in Sec. V.A.

B. Qualitative Trends in Ellipsometric Effects Measured During Interfacial Processes

Having considered in the preceding section the range of optical properties expected for different types of surface films formed in electrochemical systems, it would be helpful at this point to examine the qualitative response of ellipsometric and reflectometric parameters to the growth of certain characteristic types of surface films on metallic substrates. Such a qualitative "feel" is obviously not a replacement for, but rather the first step in the exact analysis of the interfacial phenomena detected by the ellipsometer. Also, such "feel" provides a basis for the initial guess of the optical properties of the film, required as input for a computer search of the best quantitative fit to the experimental results (see Sec. II.D). The most common form of presentation of ellipsometric effects brought about by film growth is a plot of $\delta\Psi$ vs. $\delta\Delta$. Figure 9 shows $\delta\Psi$ vs. $\delta\Delta$ plots simulated for the growth of thick transparent ($k_F = 0$) surface layers of three different refractive indices on a metal (Pt) substrate. This figure demonstrates that the $\delta\Psi$ vs. $\delta\Delta$ plot for a thick transparent layer has the form of a closed loop (progressing clockwise with the increasing film thickness). From Eq. (5) it can be

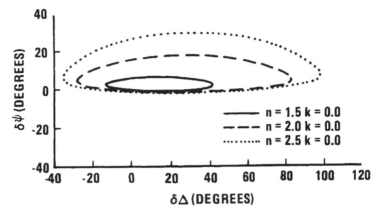

FIG. 9. Simulated $\delta\Psi$ vs. $\delta\Delta$ plots for the growth of some thick transparent layers on a Pt substrate, at $\lambda = 550$ nm, $\phi_1 = 60°$. The closed
loops generated by such films advance clockwise with film growth and
repeat every $d_f = \lambda_0/2n_f \cos \phi_1$.

seen that the loop should repeat itself for each additional film thickness,

d, given by $(4\pi/\lambda_0)dn_f \cos \phi_1 = 2\pi$, i.e., for each additional growth by

about $\lambda_0/2$. The qualitative feature to notice is that the dimensions of

the loop depend on n_F. The larger n_F is, the larger are the maximum

changes in both Δ and Ψ recorded during the growth of such thick

transparent layers. The dimensions of such loops are thus a qualitative

measure of n_{film} for transparent layers that grow on metallic substrates.

Even when the maximum thickness of the film grown is not sufficient to

close the loop, but is sufficient to observe the first inflection, which

corresponds typically to thicknesses of less than 50 nm, the inflection

point can serve as a qualitative measure for n_f. Relevant electrochemi-

cal problems are the growth of relatively thick transparent oxide or

polymeric films on metal electrodes. It is clear that a qualitative distinc-

tion between a transparent dense oxide layer ($n_f = 2.2-2.5$) and a

transparent heavily hydrated oxide or hydroxide layer ($n_f = 1.4-1.5$)

can be easily made for sufficiently thick films, according to the size of

the $\delta\Psi$ vs. $\delta\Delta$ loop. Figure 10 demonstrates the effect of finite absorp-

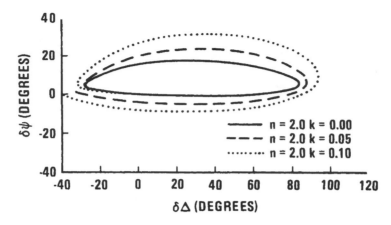

FIG. 10. Effect of finite absorption on some simulated $\delta\Psi$ vs. $\delta\Delta$ plots for film thicknesses 0–300 nm. Conditions as in Fig. 9.

tion, i.e., $k_f \neq 0$, on the plot for thick films. Instead of the repeating closed loop obtained when $k_f = 0$, the introduction of relatively small values of k_f generates an expanding spiral. (More complicated contours may be obtained for $\phi_1 > 70°$.) The extent of the opening of the spiral per cycle reflects the deviation from $k_f = 0$. Similar qualitative criteria for n_F and k_F are somewhat less straightforward for thinner films. Figure 11 demonstrates a situation parallel to that described in Fig. 9, but the film thickness is limited to 5 nm. For such thin transparent films the effect on Δ is roughly proportional to $(n_f - n_{soln})d$, and is thus largest for a given film thickness when n_f is largest. The curves are well distinguished as long as complete transparency ($k_f = 0$) is assumed. In reality, however, this assumption cannot be safely made under any circumstances, and thus a slightly absorbing film can be mistaken for a completely transparent film with a larger n_F and with a different thickness. Such interplay between n, k, and d for very thin films (typically, <10 nm), can be resolved in many cases by the addition of reflectometric measurements, as demonstrated below.

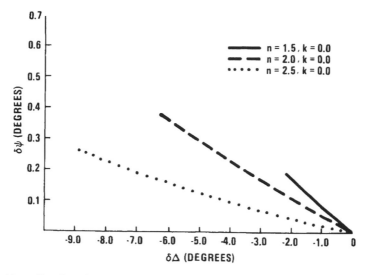

FIG. 11. Simulated $\delta\Psi$ vs. $\delta\Delta$ plots for the growth of transparent films on a metal substrate, plotted for thicknesses 0—5 nm. Conditions as in Fig. 9.

Once k_f reaches a value of about 0.05 or larger, the qualitative distinction between transparent and absorbing thin films becomes easier. This is shown in Fig. 12. It is seen from this figure that the qualitative difference is that absorbing films with $k_f = 0.05$ or above, growing on metal substrates, generate negative changes in Ψ for angles of incidence $\phi_1 = 60$—$70°$, but thin films with $n_f > n_{soln}$ that are completely transparent or very slightly absorbing ($k_f < 0.05$) exhibit positive changes in ψ with film growth. This special behavior of transparent thin films originates from the dominance of interference effects, while for absorbing films the effect of energy loss dominates in the determination of the sign of $\delta\Psi$. This special qualitative behavior, i.e., $\delta\Psi > 0$ at high angles of incidence, can be used for the qualitative identification of the growth of thin transparent films in electrochemical systems. The special effect of the growth of such a transparent film on Ψ is expected to vanish at angles of incidence closer to $45°$. This dependence on ϕ_1

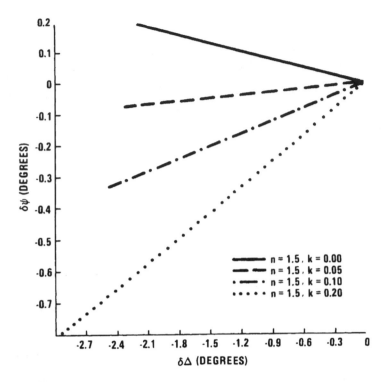

FIG. 12. Simulated δΨ vs. δΔ plots for the growth of thin films with different degrees of absorption, plotted for thicknesses 0–5 nm. Conditions as in Fig. 9.

serves as a further check on such identification (see Ref. 25). Figure 12 also demonstrates the different nature of the responses of Δ and Ψ to the growth of a film with given optical properties. As an "amplitude parameter," Ψ has a high sensitivity for energy loss in the film and thus responds sensitively to variations of k in films of given n and d. On the other hand, the phase parameter Δ serves essentially as a "thickness gauge," reflecting the overall film growth with only slight sensitivity to the value of k_f. Thinking of Δ as a "thickness gauge" and of Ψ as an "energy-loss parameter" is a useful qualitative concept for thin films.

The final example of qualitative evaluation is brought in Fig. 13(a). This is a case of two highly absorbing films. The $\delta\Psi$ vs. $\delta\Delta$ plot for such films is typified by the absence of repeated loops. The plot will rather converge upon a single point when the film thickness is above about 50—100 nm. This is so because as the film thickness exceeds the penetration depth of the light probe, the optical behavior of the filmed substrate becomes identical to that of two semi-infinite adjacent phases (solution film), corresponding to a constant (Ψ, Δ) point. In addition to the qualitative behavior of highly absorbing phases, it is also important to notice from Fig. 13(a) and (b) that $\delta\Psi$ vs. $\delta\Delta$ plots for two films with very different optical properties could become practically indistinguishable for thickness $\leq$ 12 nm. This is a demonstration of the origin of possible errors in the qualitative identification of thin films according to ellipsometric measurements. The film with n = 1.43—1.45i corresponds to measured properties for a roughened metal alloy surface [44], and the film with $\hat{n}$ = 2.14—0.77i could correspond to a strongly doped dense oxide layer. Each of these films could arise, in principle, on the same metal electrode surface being examined in a combined ellipsometric-electrochemical investigation. In fact, misinterpretation of oxide film growth as a process of metal roughening resulting from such indistinguishable $\delta\Psi$ vs. $\delta\Delta$ plots has occurred in at least one previous investigation [45]. Is there a way out? The answer is yes. First, if the surface film can be further grown (d > 12 nm), the $\delta\Psi$ vs. $\delta\Delta$ plots become easily distinguishable [Fig. 13(a)]. If the surface layer cannot be further grown, a distinction can still be made by adding reflectometric measurements, as discussed in Sec. II and shown in Fig. 13(c). One last hint is that one should never forget the electrochemical information available. The distinction between two very different surface processes, such as oxide growth and metal roughening, can be aided in most cases by the electrochemical behavior of the interface.

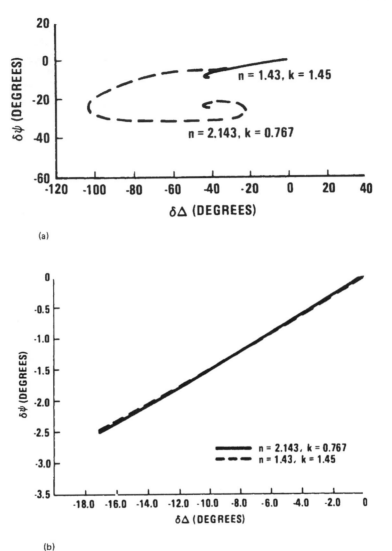

(a)

(b)

FIG. 13.   (a) δΨ vs. δΔ plots for two highly absorbing films on a metal substrate.   Conditions as for Fig. 9.   (b) Same as (a), but plotted at a higher sensitivity for thicknesses 0–5 nm.   (c) δR/R vs. δΔ plots for the two films which give the indistinguishable δΨ vs. δΔ plots shown in (b).   Curves from top to bottom are: $(δR/R)_S$ vs. δΔ for $\hat{n} = 1.43-1.45i$; $(δR/R)_S$ vs. δΔ for $\hat{n} = 2.143-0.767i$; $(δR/R)_p$ vs. δΔ for $\hat{n} =$

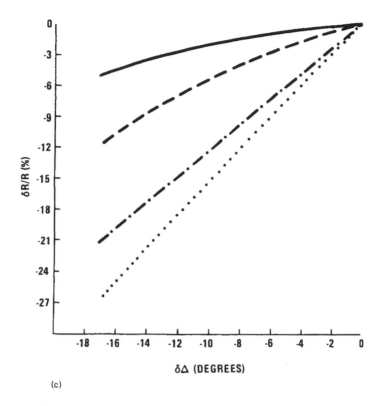

(c)

FIG. 13. Continued: 1.34–1.45i; $(\delta R/R)_p$ vs. $\delta\Delta$ for $\hat{n}$ = 2.143–
0.076i. All plots are for d_f = 0–5 nm.

V.   EXPERIMENTAL RESULTS

A.   Oxide and Polymeric Films

1.   *Determination of Film Thickness*

Ellipsometric investigations of films formed in electrochemical systems
often seek the film thickness and the rate low for film growth. The
determination of the thickness of a surface film in situ by a nonelectro-
chemical technique can be very valuable for complete characterization of
the film. When the faradaic efficiency for film growth is different than
1 and/or the stoichiometry and the degree of porosity of the film are
unknown, the thickness cannot be deduced from coulometric data. In

fact, in very few cases are the faradaic efficiency or stoichiometry known a priori with a high degree of certainty. For example, ellipsometric measurements showed that even when most of the growth of a thick oxide film takes place at 100% faradiac efficiency, early stages in the growth can deviate from this pattern. Thus, in the case of oxide growth on molybdenum in acetic acid [46], ellipsometric measurements during the early stages of galvanostatic growth demonstrate that film growth first lags behind the increase in voltage, but then it readjusts. Such behavior can be due to a dissolution-precipitation mechanism [46] or can be due to parallel processes of film growth and metal dissolution where the rate of the latter decays after a certain minimum film thickness passivates the surface.

In ellipsometric investigations of the electrochemical growth of thick films, it has been reported in quite a few cases that the optical properties of the film do not vary with film thickness. Such behavior can be verified from the nature of the Ψ vs. Δ plot obtained during thick-film growth. Examples of this behavior include dense oxide films on Ta [47], Mo [46], or V [48]; porous hydrous oxide layers on Ir [11] or Ni [49]; and polymeric layers such as poly(phenol) [50]. Computer fitting of the experimental results to a model that assumes the growth of a single uniform layer [see Eq. (20)] yields good fits under such circumstances (see Fig. 14). The film thickness at each stage of growth can be derived from analysis of the ellipsometric data and can be used, in turn, to probe the effect of layer thickness on various physical or electrochemical properties. For example, the overall voltage drop across a barrier oxide film required to pass a given ionic current of growth has been found for vanadium oxide to vary exactly linearly with the ellipsometrically determined film thickness [48], as shown in Fig. 15. Such behavior is typical for growth driven by the electric field across the oxide layer.

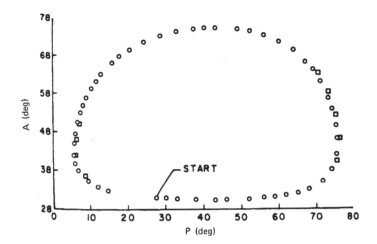

FIG. 14. Ellipsometer null readings for the growth of an anodic oxide film on vanadium at a current density of 275 μA cm^{-2}, in a solution of acetic acid containing small amounts of water and $Na_2B_4O_7 \cdot 10H_2O$. Open circles, first loop; open squares, second loop. The data were fitted for uniform growth of a film with $n_f = 2.392-0i$, on a substrate with $\hat{n} = 4.346-3.744i$ ($\lambda = 633$ nm). (From Ref. 48. Reprinted by permission of the publisher, The Electrochemical Society, Inc.)

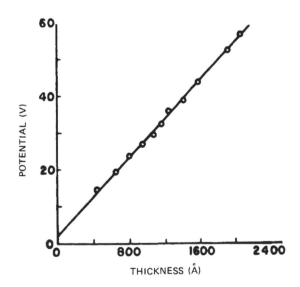

FIG. 15. The potential required for vanadium oxide film growth at a rate of 275 μA cm^{-2} vs. the optically determined film thickness. (From Ref. 48. Reprinted by permission of the publisher, The Electrochemical Society, Inc.)

In contrast, the growth of porous films usually requires much lower fields and may be accompanied by other parallel faradaic processes, such as electrodissolution.   Cases that have been investigated with ellipsometry include both oxide and polymeric films.   Hydrous oxide films are known to form on various metals by potential multicycling or by multipulsing between appropriate anodic and cathodic limits in aqueous electrolytes. Film thicknesses derived ellipsometrically during the growth of such hydrous oxide layers have been reported for nickel [49], iridium [11, 51], ruthenium [19], and iron electrodes [53].   Ellipsometric measurements showed that in all the foregoing cases, such films conformed quite well to a uniform-growth model, with thicknesses typically between 100 and 200 nm.   An example is given in Fig. 2 for a hydrous oxide and in Fig. 16 for a polymeric layer of poly(phenol) grown on a Pt electrode.

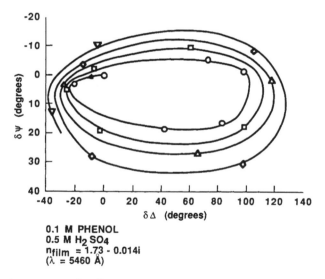

0.1 M PHENOL
0.5 M H_2 SO_4
n_{film} = 1.73 - 0.014i
(λ = 5460 Å)

FIG. 16. $\delta\Psi$ vs. $\delta\Delta$ plot for the anodic growth of a polymeric layer at a constant applied potential from a solution of 0.1 M phenol in 0.5 M H_2SO_4.   Thanks to the low value of k_f, film growth could be monitored ellipsometrically through more than five loops.   Each complete loop corresponds to ca. 230 nm of film thickness.   Computer fitting assuming uniform film growth gave the curve presented in the figure, corresponding to $\hat{n}$ = 1.73—0.014i at λ = 546 nm.   (From Ref. 50.   Reprinted with permission from North-Holland Physics Publishing, Amsterdam.)

The reasonable fits to a model which assumes a uniform film, suggest that the factors that determine the effective optical properties in such composite films, i.e., the volume fractions of the individual phases and the microstructure within the composite layer, do not vary significantly with film thickness. Combined with the results of quantitative elemental analysis for similar hydrous oxide films [54], ellipsometric film-thickness measurements [11] showed that such films may have a volume density as low as 20% of the compact oxide. Such results thus provide direct evidence for a high degree of porosity in a film.

Some hydrous oxide films, as well as some conducting polymer films, are associated with a large charge capacity, which may reach values as high as 100 mF cm^{-2} (geo.), depending on the overall film thickness. This charge capacity depends on the overall number of sites within the film that are active in the redox process involving injection of electrons and ions into or out of the film (see further discussion in Sec. V.C). In these cases, thickness determination by ellipsometry provides inportant information on the number of charging centers per unit volume of the film. This has been demonstrated recently for the case of films of conducting polymers [55,56]. For example, a film of polyaniline was found, from combined voltammetric and ellipsometric measurements, to be associated with a charge density of 840 C cm^{-3} [56]. This result is demonstrated in Fig. 17. The concentration of active redox centers derived from this charge density is 1×10^{22} cm^{-3} (assuming 1e exchanged per site), which is about four times larger than that reported on the basis of similar ellipsometric measurements for a film of polyvinyl ferrocene [55]. The volume density of redox centers derived with the aid of ellipsometry should aid in the understanding of the mechanism of charge transport in this important class of compounds. In the last work mentioned [55], ellipsometric evaluation of film thicknesses for PVF in its reduced and oxidized forms revealed that the oxidized form of the film was thicker by about 15%, a change interpreted as being due to film

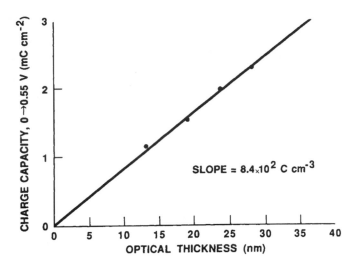

FIG. 17. Voltammetric charge capacity (measured at 50 mV sec^{-1}) against ellipsometrically determined film thickness for a film of polyaniline grown in aqueous 2 M HCl.

swelling from the uptake of counterions and attendant solvent molecules during the film oxidation process. A promising future combination of in situ techniques is that of ellipsometry and the quartz crystal micro-balance (QCM) technique [57]. A combination of these two in situ measurements should yield film thickness variations (ellipsometry) asso-ciated with the uptake of a given mass of solvent molecules or ionic species (QCM) during an electrochemical film-conversion process.

For some polymeric films it has been shown that rate laws for electro-chemical film growth can be derived from ellipsometric measurements taken under potentiostatic conditions. As shown in Fig. 18, for the anodic growth of a polymeric film on a Pt electrode from an aqueous solution of 1-naphthol, a parabolic rate law was found, i.e., the thick-ness increased with the square root of time [50]. The rate of growth was, however, found to increase with the applied potential. This behavior was interpreted as being due to film growth being controlled by field-assisted electron transport through the growing polymeric

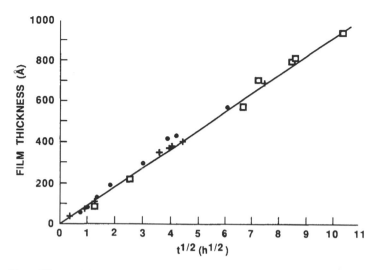

FIG. 18. Film thickness derived ellipsometrically (uniform film growth model) vs. time$^{1/2}$, for the growth of a polymeric layer from a 1 mM solution of naphthol in 0.5 M H_2SO_4 at a constant applied potential of 0.75 V. The points are from independent measurements at three different wavelengths. (From Ref. 50. Reprinted with permission from North-Holland Physics Publishing, Amsterdam.)

layer [50]. The overall thickness of the polynaphthol film grown under such conditions was found to be below 100 nm. On the other hand, for polymeric films grown anodically from aqueous solutions of phenol on Pt electrodes, the rate law for film growth derived from ellipsometric measurements was different: Film thickness increased linearly with time [50]. This is shown in Fig. 19. In the last case, film thicknesses as large as 1 µm could be reached, and the complete process of growth of such thick layers could be followed ellipsometrically. The rate law found in this last case shows that growth of the polymeric film is not limited by transport through the thick (and porous) film. Based on the measured dependence of growth rate on applied potential (Fig. 19), it was suggested that the growth of the polyphenol film was controlled by electron transfer through a very thin polymeric barrier layer [50]. Similar results have been reported very recently for the growth of polymeric

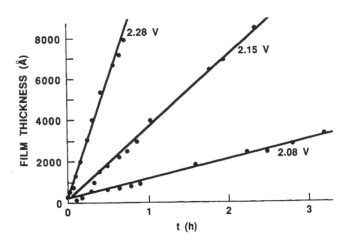

FIG. 19.   Film thickness derived ellipsometrically (uniform film growth model) vs. time, for the growth of a polymeric layer from 0.1 M phenol, 0.5 M H_2SO_4 solution, at three different constant applied potentials. (From Ref. 50.   Reprinted with permission from North-Holland Physics Publishing, Amsterdam.)

layers deposited potentiostatically from solutions of thionine and of thiopene [58].   The polythionine layer was limited in thickness to 5 nm and was shown ellipsometrically to exhibit a parabolic growth rate law. The polythiophene layer was grown to thicknesses of 80 nm.   Following an initial stage of instantaneous growth to about 35 nm (apparently nucleation), the film exhibited a linear growth rate law.

2.   Oxide and Polymeric Films: Optical Properties

In the case of the controlled growth of a thick uniform film, the determination of optical properties from ellipsometric measurement is straightforward.   Such properties are evaluated with no assumptions other than that of film uniformity, which can be tested by the quality of the fit · obtained, as explained in Sec. II.D.   Excellent fits have been obtained for the growth of thick oxide films on valve metals, e.g., tantalum [47], niobium [59], zirconium [60], as well as on vanadium [48] or molybdenum [46].   In all these cases, the refractive index of the anodically grown film in the visible region is real and is found to be relatively high, varying typically between 2.10 and 2.35. Such a high value of n_f has been found also for thinner ($\leq$ 4 nm) oxide films formed on titanium in acid solutions, a result which proved that such thin films are already close in stoichiometry and density to bulk (rutile) TiO_2 (see Ref. 61). In strong contrast to such dense oxide films that grow by a high-field mechanism, ellipsometric measurements of oxide films formed by potential multipulsing or multicycling show much lower values for the real part of the refractive index. A value of n = 1.46 was reported from ellipsometric measurements on films formed on Ni electrodes by potential multicycling in 5 M KOH [49], and a practically identical value was reported for the thick oxide layer formed on iridium electrodes by multicycling in acid solutions [11]. An even lower value of n_f = 1.41 was reported for the oxidized form of the thick hydrous oxide layer formed on iron electrodes by potential cycling in air-saturated 0.05 M NaOH [53]. These values of n_f are close to n of the electrolyte solution, which typically varies in the visible region between 1.33 and 1.36. A film with optical polarizability closer to that of the solvent than to that of the corresponding crystalline oxide is optical evidence for low density in a porous layer with a high water content. Some electrochemical characteristics of such films, such as their large charge capacity, which has to be associated with a high density of sites accessible by both electrons and protons, may be considered as indirect evidence for the high porosity and water content. However, the low levels of n_f revealed by ellipsometric measurements is direct in situ evidence about the nature of such layers. The examples above suggest that the real part of the complex refractive index, as determined from ellipsometry, can serve as a general measure for film density or for film porosity. However, estimates of density according to the real part of n_f alone are restricted to transparent or very slightly absorbing films (k_{film} < 0.1). The reason for this is that, in absorbing films, the value of the real part of n_f may be significantly affected by modes of energy loss in the film, as can be realized from the variations of n near the peak of k in Fig. 8. Immediate conclusions on film density, according to the value of the real part of n_f, are thus safe only for films with k_f equal or close to zero in a wide λ region. This requirement is fulfilled for the porous oxide films mentioned above if under appropriate applied bias (for the hydrous Ir oxide or Ni oxide films when under cathodic applied bias).

The Ψ vs. Δ plots for the growth of thick films with significant absorption are different from those produced by transparent or semitransparent films (see Fig. 10), and a thick absorbing film is easily recognized by the qualitative nature of the ellipsometric growth characteristic. An example of the different $\delta\Psi$ vs. $\delta\Delta$ plots obtained for colorless and for colored forms of the same oxide film are given in Fig. 20. Values of k_f close to 0.10 were reported for the colored anodic forms of hydrous Ir oxide [11] and hydrous Ni oxide [49], and k_f of 0.33 and 0.47 were

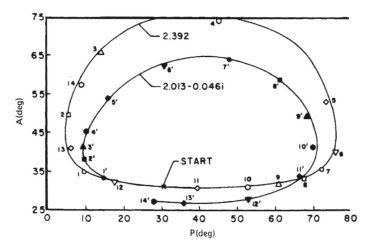

FIG. 20. Ellipsometric null readings for the colorless anodic form (open symbols) and for the colored cathodic form (filled symbols) of anodic vanadium oxide. Identical numbers on the two different plots correspond to the same stage of oxide growth. Fitting was obtained for $n_f = 2.332 - 0i$ (colorless) and $\hat{n}_f = 2.013 - 0.046i$ (colored). ($\lambda = 633$ nm.) (From Ref. 48. Reprinted by permission of The Electrochemical Society, Inc.)

reported for the colored forms of vanadium [48] and tungsten oxide [60] films, grown galvanostatically and subsequently cathodically converted. The value of k_f at a given λ in the colored forms of such films is determined not only be the proximity to an absorption peak (or edge) and by the optical transition probability but also by the density of the film. This is the reason why the more porous structures of the Ir and Ni hydrous oxides exhibit lower values of k_f. The level of k_f reflects the volume density of "absorption centers" in the film and their "oscillator strength." This last description is in terms appropriate for localized electronic states. For films in which the electronic states are sufficiently delocalized, a more appropriate description of the photo-induced electronic transitions would be in terms of a band model for the solid phase. For example, for tungsten trioxide, cathodic conversion is claimed to involve the introduction of new midgap states as well as the population of a band derived from antibonding $d-\pi^*$ states. Light absorption at energies lower than the energy of the gap for transparent WO_3 (ca. 3.1 eV) may thus involve transitions to or from the midgap states as well as intraband transitions due to the electrons in the $d-\pi^*$ band. To differentiate between such types of electronic transitions, the spectrum of k_f is required (see Sec. IV.A). From the magnitude of k_f at a single wavelength, it can only be stated that the levels of k_f reported for the colored forms of the oxide films (0.1–0.5) are typical for transitions of high probability between localized electronic states as well as for direct interband transitions. The origin and level of k_f may be also microstructure related, as discussed in Sec. V.C.

Several cases have been reported in which two kinds of oxide films, usually a superficial compact layer covered by a more open-structured and hydrous layer, were detected and characterized by ellipsometric measurements. These ellipsometric measurements demonstrated the different optical parameters and thicknesses of the two different types of oxides. For example, Shibata and Sumino [62] showed that the

passivity of Pt electrodes can be broken to some extent in 0.5 M H_2SO_4
at high potentials of 2.1–2.25 V vs. RHE. This results in the growth
of a thicker oxide layer on top of the superficial thin oxide layer (1 nm)
that grows on Pt electrodes between about 0.85 V and the onset of
oxygen evaluation. Gottesfeld et al. [63] followed this electrochemical
process ellipsometrically. Their results are shown in Fig. 21. They
found that the thicker oxide layer that grows at higher potentials reaches
a typical thickness of 8–9 nm and has a refractive index of $\hat{n}_f$ = 2.1-
0.01i. The very low extinction coefficient for this layer should be
contrasted with the apparent high value of k_f for the superficial under-
lying oxide, for which $\hat{n}$ = 3-1.5i [64]. This low value of k_f for the
thick oxide implies low electronic conductivity, which may explain the
very high overpotential for the reduction of this outer oxide layer shown
to electroreduce only within the Pt-H potential region [63]. Because of

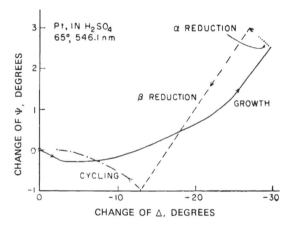

FIG. 21. Ψ vs. Δ plot for the formation and stepwise reduction of oxide
films α and β on Pt in 0.5 M H_2SO_4 (20 h growth at 2.25 V vs. RHE).
Solid curve, growth at 2.25 V. Dotted curve, reduction of the thin
"regular" Pt oxide layer, completed at about 0.5 V. Dashed curve, reduction of the thicker outer layer within the Pt-H region. The last step
produces a rough Pt surface. Dashed-dotted surve, gradual smoothing
of the rough Pt surface by multicycling 0 $\longleftrightarrow$ 1.6 V. (From Ref. 63.
Reprinted with permission from Les Editions de Physique.)

the low value of k_f and the high potential of film formation, involving
the apparent breakdown of the passivity of the thin superficial Pt oxide,
it was suggested that the thicker external oxide film is a hydrated form
of Pt(IV) oxide [63]. Recent detailed examination by XPS [65] is in
good agreement with this earlier ellipsometric assignment. A similar
structure of an inner thin and compact oxide layer covered by an outer
porous layer was characterized by ellipsometric measurement on iron
electrodes in alkaline electrolyte by Huang and Ord [53], and their
results are shown in Fig. 22. They found $\hat{n}_f = 2.0-0.25i$ and d = 4 nm
for the compact layer, compared with $\hat{n}_f = 1.41$ and 1.38 ($k_f = 0$) for
the oxidized and reduced forms of the outer porous and hydrous layer,
respectively. The last layer was grown up to 140 nm by potential
multicycling. Similar ellipsometric results for iron in alkaline electrolytes
were reported by Albani et al. [66]. Szklarska-Smialowska et al. [141]
showed ellipsometrically that in the presence of EDTA, reversible formation and complete reduction of the thin oxide layer on Fe in NaOH
solutions take place, and the phenomenon of growth of a thick hydrous
layer during cycling is prevented.

In several cases ellipsometry was helpful in characterizing the
different nature of the oxide films formed on the same substrate under
different growth conditions. Visscher and Barendrecht tabulated different values of n_f derived for oxide films grown electrochemically on Ni
electrodes [68]. In acid solution a thin passivating layer closer in composition to NiO, with $\hat{n}_f = 2.10-0.46i$ (500 nm), has been evaluated [67],
whereas in alkaline solution, the thin oxide layer has a similar real part
of n_f, but $k_f = 0$. The passive film on nickel has been reexamined
ellipsometrically very recently by Kang and Paik [69]. In contrast to
the previous reports (e.g., Ref. 68), which were based only on Ψ
and Δ measurements and thus required various assumptions in the
analysis, Kang and Paik took combined ellipsometric and reflectometric
measurements, which allowed them to solve the film at each point

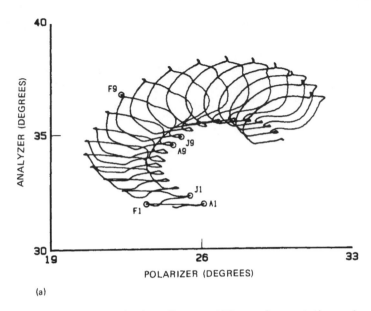

(a)

FIG. 22. (a) Ellipsometric data for repetitive galvanostatic cycles at 87 $\mu A\ cm^{-2}$ applied to an iron electrode at 0.5 M NaOH. The small loops correspond to the continuous growth and the partial reduction of a thick and very porous film on top of a thin inner oxide layer [see (b)]. (b) Analysis of the results presented in (a), in terms of an inner oxide layer 3.8 nm thick with $\hat{n}_f = 2.0\text{-}0.25i$, and a thick (140 nm) porous layer with $\hat{n}_f = 1.411\text{-}0i$ and $\hat{n}_f = 1.381\text{-}0i$ in its oxidized and reduced forms, respectively. (From Ref. 53. Reprinted by permission of the publisher, The Electrochemical Society, Inc.)

in the growth process with no assumptions made on the nature of the film (see Sec. II.C). The results reported by Kang and Paik for the passivating oxide on nickel are shown in Fig. 23. The evaluated ranges of variations in optical constants for the oxide film during its growth were: $n_f = 1.8\text{--}2.0$, $k_f = 0.2\text{--}0.4$, $d_{max} = 1.25$ nm. These results are in good accordance with information on the passive film on nickel in acid solutions obtained with other techniques [70], and they suggest that some of the conclusions obtained in early ellipsometric work on such thin passivating films, as well as some of the resulting implications on the general nature of the passivation process [71,72], were not correct. One question that remains open, however, is whether

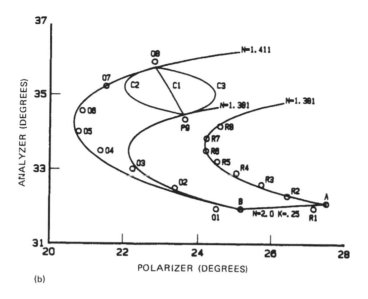

(b)

FIG. 22.   Continued

the apparent high value of k_f obtained for the thin Ni oxide layer in
acid solutions [69] (but not in alkali!) is not a result of simultaneous
metal surface roughening and oxide film growth.   A change in ellipsometric parameters due to some substrate roughening can result in
apparent absorption in the film evaluated if a single oxide film is assumed
in the analysis.   A possible check for metal surface roughening is by
verifying that the electroreduction of the oxide film restores the initial
ellipsometric parameters of the "bare" substrate.   Obviously, this is
possible only for fully electroreducible oxide layers.   Thicker film
growth on Ni by potential multicycling leads to films with a much lower
$\hat{n}_f$ (1.44-0i at 632.8 nm), suggesting a much lower density and higher
H_2O (or OH) content, and leading to film assignment as β-Ni(OH)$_2$ [49].

For relatively thick anodic films on Ru, formed in acid solutions,
Rishpon and Gottesfeld [19] showed from ellipsometric measurements that
the results conform to a uniform film-growth mode, where $\hat{n}_f$ = 3.53-
0.936i is found for oxide film growth at constant potential (with a

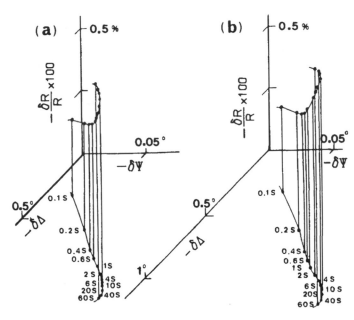

FIG. 23.   Variation of Δ, Ψ, and R with formation of a passivation film after applying a 900 mV potential jump to a nickel electrode.   The time elapsed after the potential jump is marked along the curve.   (a) pH 3.21; (b) pH 1.81.   (From Ref. 69.   Reprinted with permission from North-Holland Physics Publishing, Amsterdam.)

maximum thickness of about 30 nm), and $\hat{n}_f = 1.73-0.47i$ (and a thickness exceeding 100 nm) is found for films grown on Ru by potential multi-cycling in the same solution of 0.5 M H_2SO_4.   These results are presented in Fig. 24.   Again, potential multiucycling or multipulsing is shown by ellipsometry to generate thicker films of a more open structure and a higher water content.   Growth at a constant anodic potential yielded, in this case, an oxide film with optical properties very similar to those reported for the superficial layer grown on Ru by thermal oxidation [73].

Optical properties derived for thick polymeric films provide informa-tion of a similar nature: The real part of $\hat{n}_f$ supplies information on the density of the film, and the level and spectrum of k_f reflect the nature of electronic transitions leading to absorption in the UV-visible-near IR domains.   Values of the real part, n_f, reported for electrochemically

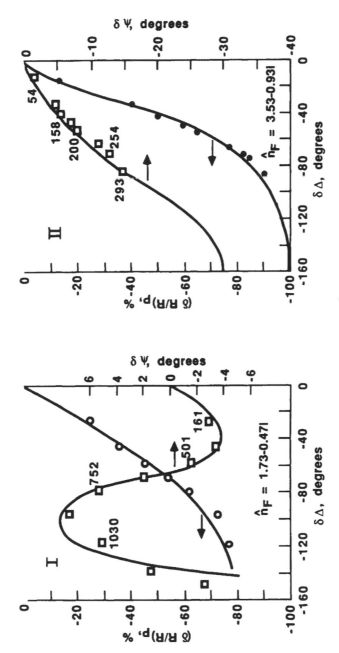

FIG. 24. Experimental points and computer-fitted curves for a uniform-film-growth model, for the growth of anodic Ru oxide films on Ru in 0.5 M H_2SO_4: I, by cyclic multipulsing at 0.5 Hz between 1.225 and −0.3 V; II, at a constant potential of 1.225 V. Thickness in angstroms designated. The ellipsometric measurements were taken at 546 nm. (From Ref. 19. Reprinted by permission of the publisher, The Electrochemical Society, Inc.)

formed polymeric films were between 1.70 and 1.74 (546 nm) for polyphenol and polynaphthol [50] (see Figs. 16, 18, and 19), 1.63 for
polybipyrazine, 1.50 for polyvinyl ferrocene (546 nm) [55], and 1.55–
1.60 for polyaniline (450–650 nm) [56]. It was suggested by Babai and
Gottesfeld [50] that an estimate of the refractive index expected for a
dense polymeric film can be obtained from the observation that bulk
homopolymers have a refractive index typically 0.05–0.10 unit larger
than that of the liquid monomer [74]. For the extinction coefficient,
k_f, the following levels were reported for various conducting polymer
films: Babai and Gottesfeld reported k_f = 0.12 for polynaphthol, k_f =
0.06 for polyphenol films grown at low anodic potentials, and k_f = 0.01
for thicker films of polyphenol grown under high anodic potentials (all
at 546 nm). The much lower value of k_f for films grown at higher
potentials was ascribed to partial breakdown of the long-range conjuga-
tion in the polymer, due to excessive oxidation [50]. Arwin et al. [75]
evaluated from spectroellipsometric measurements the absorption spectrum
of polypyrrole in the λ range 1.5–6 eV. Their results are given in
Fig. 25 and demonstrate examples of spectra of surface layers on ordi-
nary bulk electrodes obtainable by spectroellipsometry. They found an
absorption coefficient ranging from 5 to 8.10^4 cm^{-1} in the visible
domain, which corresponds to k_f of the order of 0.2. Neither the
relatively modest magnitude of k_f nor the spectrum of polypyrrole shown
in Fig. 25(a) suggests metallic conductivity in this film (see Sec. IV.A).
Gottesfeld et al. [56,76] have assumed in their analysis uniform film
growth for a layer of polyaniline grown galvanostatically on Pt in aqueous
HCl. They found real parts of n_f ranging between 1.50 and 1.60 and
extinction coefficients between 0.01 and 0.03 for the cathodic ("bleached")
form of the polyaniline films (450–650 nm). Following anodic film conversion, k_f increased up to 0.25 at 650 nm, the longest λ employed.
The value of n_f was found to decrease consistently under anodic biases,
dropping to about 1.15 at 650 nm. The results and the fit obtained for

the two forms of the film at 600 nm are shown in Fig. 26. These authors explained the lowering in the real part of n_f following film coloration by the Lorentz oscillator model, presented schematically in Fig. 8. This model predicts a lowering of the real part of n on the high-photon-energy side of a developing optical oscillator. Such lowering of n_f has to be considered together with the increase of k_f in the analysis of electrochemical film conversion, discussed in more detail in the next section.

In their recent work on the growth of polythiophene films [58], Hamnett and Hillman showed a very similar trend of a strong lowering in the real part of n_f as the film is anodically colored, e.g., n_f = 1.8 for the bleached form and 1.3 for the colored form at λ = 633 nm. These authors, employing combined ellipsometric-reflectometric measurements, evaluated variations in k_f as well as in n_f and film thickness, during the potentiostatic growth of polythiophene. At 633 nm they found k_f to increase gradually during potentiostatic growth, starting from k_f = 0.05 and approaching asymptotically k_f = 0.30 after about 5 sec. The polythiophene film grew linearly with time within the same time domain. Bouziem et al. [130] described an ellipsometric study of the cathodic electropolymerization of acrylonitrile on a nickel substrate. They found a complex behavior, and reported metallic properties in the range 450—650 nm (ε' < 0!) for films about 20 nm thick. Chao et al. reported on spectroellipsometric investigations of electropolymerized films of polymethylthiophene [131]. They ascribed features in the ε'' spectrum evaluated for this film to electronic transitions calculated by Bredas et al. [132].

3. *Electrochemical Film-Conversion Processes*

Electrochemical film-conversion processes are frequently associated with significant variations in the optical properties of the film. For films such as the oxides of tungsten [60], molybdenum [46], iridium [11,77], or rhodium [78], as well as for polymeric films such as polyaniline

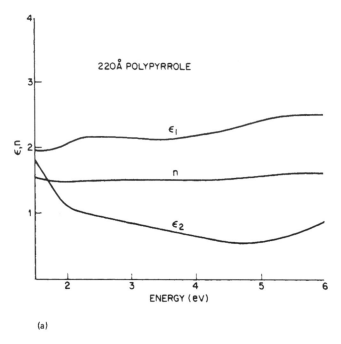

(a)

FIG. 25. (a) Dielectric response of a 22-nm film of polypyrrole grown
on a gold electrode and the corresponding real part of n_f, derived from
spectroellipsometric measurements. (b) Absorption coefficient (α) spectrum of polypyrrole calculated from the ε spectrum in (a). Data for a
thick (and apparently less compact) film are also shown. (c) Dielectric
response of poly-N-methylpyrrole grown on two different electrode
substrates. (From Ref. 75. Reprinted with permission from Elsevier
Sequoia S.S., Laussane, Switzerland.)

[76,79] or polythiophene [58], the variations in light absorption due to

such electrochemical conversion processes are sufficiently large and

reversible as to make them atractive candidate materials for display

devices. The electrochemical conversion process involves injection (or

ejection) of electronic charge and of counter ionic charge into or out of

the film (see, e.g., Refs. 77 and 78). The electrochromic activity of

the film thus depends on the extent of and the rates of both electronic

and ionic transfer and transport into and through the bulk of the film.

Ellipsometric measurements of such processes supply valuable information.

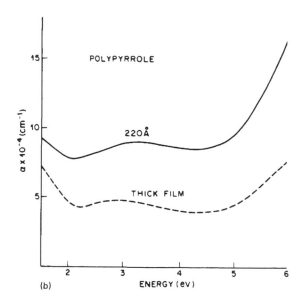

(b)

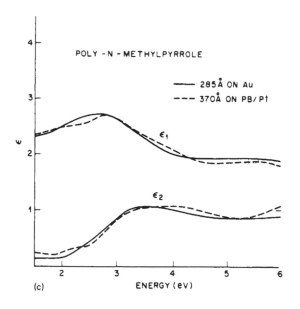

(c)

FIG. 25. Continued

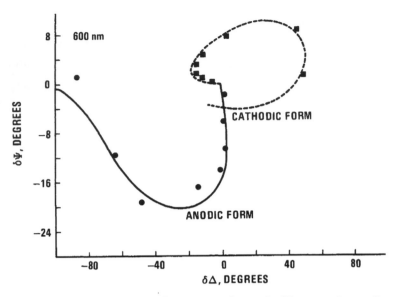

FIG. 26. Computer fitting of the experimental ellipsometric readings
recorded during galvanostatic growth of polyaniline films to a uniform-
film-growth model. Growth was from an aqueous 2 M HCl + 1 M aniline.
The curves correspond to n_f = 1.56; k_f = 0.03 for the cathodic form of
polyaniline (OV vs. Ag wire) and to n_f = 1.17; k_f = 0.20 for the anodic
form (0.52 V).

As discussed in Sec. V.A.2, comparison of optical properties and
thicknesses evaluated for a given film when under different constant
applied potentials reveals how $\hat{n}_f$ and/or the thickness of the film vary
with the applied potential. When an automatic ellipsometer is employed,
the continuous variations of the ellipsometric parameters can be recroded
while the potential is linearly scanned between the limits corresponding
to the "bleached" and "colored" forms of the electrochromic film. The
resulting "ellipsometric conversion curve" contains information on the
mode of propagation of the coloration through the volume of the film, as
demonstrated in the following examples.

DeSmet, Ord, and co-workers have described, in a series of papers,
results of ellipsometric measurements of growth and electrochemical film
conversion processes in metal oxide films. For vanadium, it was found

that following the galvanostatic growth of a transparent dense oxide
(d $\leq$ 200 nm), and with n_f = 2.392-0i (λ = 632.8 nm), the film could be
converted by the passage of a constant cathodic current to a "cathodic
form," with n_f = 2.013-0.046i [48]. For molybdenum [80], a similar
experiment gave the results 2.148-0i for the as-grown oxide and 1.969-
0.471i for the cathodically converted film (632.8 nm). Clayton and
DeSmet [48] and DeSmet [80] first showed in their work on the cathodic
conversion processes in vanadium and molybdenum oxide films that from
the Ψ vs. Δ curves recorded during the electrochemical conversion process, it could be shown conclusively that the cathodic coloration propa-
gated from the electrolyte-film to the film-metal interface. This mode of
coloration-propagation is expected for films with low rates of interfacial
ionic transfer and/or bulk ionic transport. An example of their experimental and simulated "ellipsometric conversion curves" is given in Fig.
27.

    Theoretical Ψ vs. Δ curves for various modes of film conversion can
be simulated if the optical properties of the fully colored and fully
bleached forms of the film are accurately known. For example, for
color propagation from the film-electrolyte interface, this simulation is
done by solving the variations in the ellipsometric parameters for a
system of a substrate covered by a dual film, in which the thickness of
the outer colored part increases gradually (e.g., in 100 steps) at the
expense of the inner bleached part. This is shown schematically in
mode B of Fig. 28. The optical properties of the colored part and of
the bleached part of the film assumed in the simulation of the conversion
process are the ones solved from ellipsometric readings during stepwise
film growth. The overall film thickness is assumed constant in the
simulation, unless significant film swelling (or contraction) is found from
analysis of the film growth data between the fully bleached and fully
colored forms. In the last case, the thickness variations also have to
be included in the simulation of the conversion process. The second

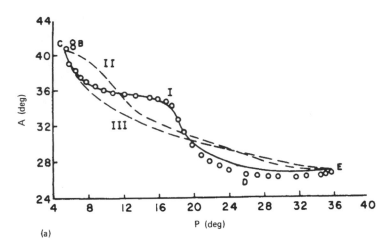

(a)

FIG. 27. (a) Ellipsometric null readings for the continuous conversion of an anodic film on vanadium with an initial thickness of 193.8 nm to a cathodic form with a thickness of 186.6 nm. $(\hat{n}_f)_{an}$ = 2.392-0i; $(\hat{n}_f)_{cath}$ = 2.013-0.046i (λ = 633 nm). Simulated curve I corresponds to mode B of conversion in Fig. 28, curve II to mode A in the same figure, and curve III to mode C. (b) The null readings for the reconversion of the cathodic form of the vanadium oxide film to the anodic form. (From Ref. 48. Reprinted by permission of the publisher, The Electrochemical Society, Inc.)

mode of propagation of coloration considered by Clayton and DeSmet was from the metal-film interface to the film-solution interface, and the third was uniform development of coloration throughout the film. These mechanisms are described schematically in modes A and C of Fig. 28, respectively. The second mechanism (Fig. 28a) is expected if electronic charge transfer into the film, or electronic transport through the film, is rate limiting in the overall charge injection process. Uniform coloration (Fig. 28c) is expected when the electronic and ionic processes within the film have very similar rates, or when both are very fast. Simulation of the ellipsometric conversion curves for these three cases reveals that the distinction between them can be made very readily (Fig. 27). This analysis of the mode of spatial propagation of a charging process within a surface film is a unique capability of ellipsometric measurements.

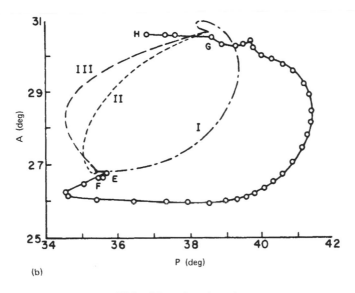

(b)

FIG. 27.   Continued

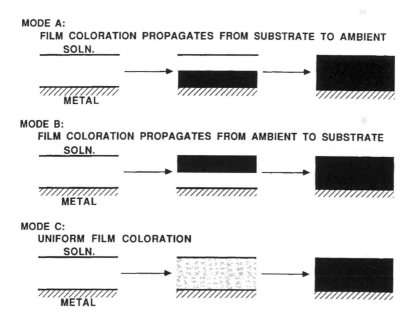

MODE A:
   FILM COLORATION PROPAGATES FROM SUBSTRATE TO AMBIENT

MODE B:
   FILM COLORATION PROPAGATES FROM AMBIENT TO SUBSTRATE

MODE C:
   UNIFORM FILM COLORATION

FIG. 28.   The three possible modes of electrochemical film conversion that can be differentiated by ellipsometric measurements.   Mode A corresponds to a process of charge injection limited by electron transport, mode B to a rate limited by insertion from solution, and mode C to a case of high levels of both ionic and electronic transport rates within the film.   (According to Refs. 48 and 80.)

For dense oxide films such as anodic vanadium and molybdenum oxides, the reverse process of (aodic) bleaching has also been found from the measured ellipsometric conversion curve to propagate from the oxide-film interface into the film. Thus the film does not pass through the same intermediate optical states when it bleaches and when it colors. This results in hysteresis in the ellipsometric conversion curve, as seen from comparison of Fig. 27(a) and (b). (In their later work on molybdenum oxide films [46], DeSmet and Ord observed with the aid of the ellipsometer that the large anodic field across the growing oxide introduces a slight optical anisotropy, which is due to the electrostriction phenomena. Such optical effects in valve metal-oxide films that are of some theoretical interest are rather small in magnitude and quite insignificant when the properties of such oxide films are evaluated for purposes other than the study of electrostriction.)

The optical parameters and ellipsometric conversion curves evaluated for porous and hydrous oxide layers are found to be quite different. Hopper and Ord found that following oxidation-reduction cycles applied to a Ni electrode in 5 M KOH, the surface film formed showed $\tilde{n}_f$ = 1.46-0i for the cathodic form and $\tilde{n}_f$ = 1.6-0.117i for the anodic form [49]. Gottesfeld and Srinivasan found $\tilde{n}_f$ = 1.44-0i for the cathodic form of the hydrous oxide grown on iridium by potential multicycling in acid solutions, and for the anodic form of the same layer they found $\tilde{n}_f$ = 1.41-0.08i [11]. It is clear from these optical properties that in the last two cases, the cathodic form of the oxide is "bleached" while the anodic form is "colored." Both n_f and k_f are relatively small and similar in magnitude for the porous oxide layers formed on nickel and iridium. When electrochromic conversion takes place in the latter type of film, the process of ionic insertion is much more rapid. This is reflected in the uniform propagation of coloration (and of bleaching) through the volume of the film, as demonstrated by the ellipsometric conversion curves in the case of the hydrous oxides of nickel [49] and of iridium [51]. Under such

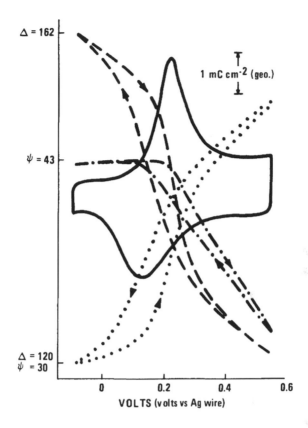

FIG. 29. Variation of current (solid curve), charge (dotted curve), Δ
(dashed curve), and Ψ (dashed-dotted curve) during a triangular cycle
at 50 mV sec^{-1} for a Pt electrode covered by a film of polyaniline about
80 nm thick, immersed in an aqueous 2 M HCl solution. Optical measurements taken at λ = 550 nm, ϕ_1 = 60°.

circumstances the ellipsometric conversion curve is associated with very
little hysteresis, because a film that converts uniformly passes through
identical intermediate optical states during coloration and during bleaching.

This last reversible form of ellipsometric conversion curves was also
demonstrated for the coloration and bleaching of films of polyaniline [76].

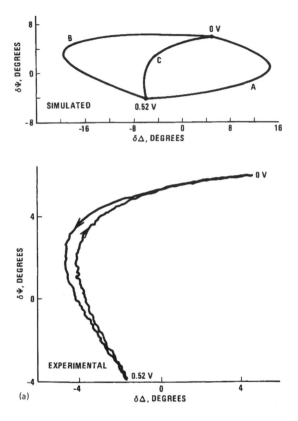

FIG. 30.   (a) Simulated and experimental ellipsometric conversion curves for the electrochemical conversion of a polyaniline film on Pt in 2 M HCl, recorded at 550 nm and 60° incidence during a 50 mV/sec scan between 0 and 0.52 V vs. Ag wire.   Optical properties of the film at 0 V: n = 1.48, k = .013; and at 0.52 V: n = 1.30, k = 0.060.   Film thickness: 115 nm.   Simulated curves A, B, and C correspond to the respective conversion modes described in Fig. 28; (b) Same as (a) for a polyaniline film thickness of 150 nm.   (From Ref. 76.   Reprinted by permission of the published, The Electrochemical Society, Inc.)

In the last case it was shown that an early step in the process of anodic coloration requires injection of an initial "triggering" charge, which may be associated with a very limited initial propagation of the coloration from the metal-film interface that is rather difficult to detect optically.

Following this initial small injection of charge, the major part of the coloration process is ellipsometrically observed to develop uniformly through the film. Figure 29 demonstrates the variations of Ψ and Δ with potential for electrochemical film conversion in a polyaniline layer about 100 nm thick, as recorded during a potential scan at a rate of 50 mV sec^{-1}. The voltammetric current and charge are also shown in the figure. The negative change of Δ recorded during anodic film conversion at 550 nm, and seen in Fig. 29, is to a large extent due to the lowering of the real part of n_f during coloration, discussed in Sec. V.B. [Both $(\delta\Delta/\delta k)_{n,d}$ and $(\delta\Delta/\delta n)_{k,d}$ are calculated to be positive for this optical system at 550 nm.] Figure 30 shows ellipsometric conversion curves for the polyaniline film, recorded at 550 nm for two film thicknesses. Compared with the simulated curves for the three possible modes of coloration-propagation, the almost hysteresis-free experimental conversion curve is seen to fit well the model of uniform propagation of coloration in the polyaniline film. This mechanism is allowed by high levels of both electronic and ionic conductivities that prevail through the larger part of the conversion process [76].

B. Submonomolecular and Monomolecular Chemisorbed Films

1. Theories and Experimental Optical Results for Chemisorption of Nonmetallic Adsorbates on Metal Substrates

In the investigation of metal-gas interfaces, ellipsometric studies of chemisorption are relatively scarce. On the other hand, in electrochemical systems the use of ellipsometry began with an emphasis on the study of electrochemisorbed layers, such as oxide and hydrogen layers, UPD metal layers, and electrosorbed organic molecules, all on the monolayer or submonolayer coverage level. The optical effects in the visible region due to the chemisorption of nonmetallic species on metal substrates are a complex combination of effects due to new electronic states formed by the interaction of the absorbed species with the metal substrate and the

perturbation of optical properties of the metal substrate surface; these
are all added to the contribution of electronic transitions in the "adsorbate itself." The resolution of these components of the overall measured
effect is very complicated in most cases. Indeed, such contributions to
the measured ellipsometric effect exist for every process of film growth
on a metal substrate. However, although these effects become quite
insignificant in the analysis of thick films, substrate perturbation effects
may dominate the overall measured ellipsometric signal in the case of thin
chemisorbed layers. Habraken et al. [81] summerized theories employed
for the analysis of optical effects due to submonomolecular chemisorbed
films on metal substrates and described attempts to use such theories to
analyze ellipsometric results for the chemisorption of oxygen from the
gas phase on Cu, Ag, and Ni single crystals. The use of a phenomenological macroscopic model of a single thin film with a uniform isotropic
complex refractive index results, in such cases, in unusual optical
properties that in no way resemble the properties of the corresponding
thick oxide. For example, for an arbitrary choice of thickness for the
chemisorbed oxygen layer as 0.4 nm per monolayer, the results at 632
nm for oxygen chemisorbed on Ag (110), Cu (111), and Ni (100) were
$\hat{n}_f = 2.7-2.0i$, $4.3-1.8i$, and $2.1-4.0i$, respectively [81]. For comparison,
for bulk Cu_2O, values have been reported of $\hat{n}_f = 2.5-0.1i$ [82], and for
bulk Ni oxide, $\hat{n}_f = 2.33-0i$ at the same wavelength [83]. The strong
dissimilarity between the optical properties derived for the submonomolecular chemisorbed layer and for a thick oxide film is obvious, as is
the "metallic nature" of the optical properties evaluated based on the
assumption of a single thin film due to the chemisorbed oxygen. This
"metallic nature" is caused by the significant modulation of the optical
properties of the metal surface brought about by the chemisorption
process. Results of very similar nature were obtained from ellipsometric
measurements of the monolayer of oxide formed electrochemically on Pt:
$\hat{n}_f = 3-1.5i$ [64], and these pseudometallic optical properties originate

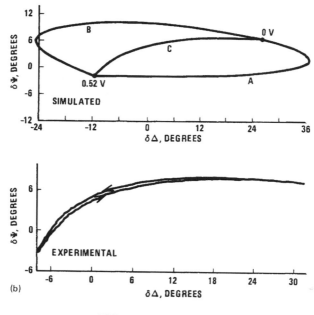

FIG. 30. Continued

most probably from the electronic effects mentioned above for oxygen chemisorption on metals. Attempts to resolve the components of the ellipsometric effects measured for a chemisorbed layer by an approximation such as

$$\delta\Delta_{total} = \delta\Delta_o + \delta\Delta_{N_F} + \delta\Delta_\tau \qquad (32a)$$

and

$$\delta\Psi_{total} = \delta\Psi_o + \delta\Psi_{N_F} + \delta\Psi_\tau \qquad (32b)$$

were described (e.g., Ref. 81), where the first term corresponds to the optical effect of the adsorbed layer itself, the second is due to the resulting perturbation of the free-electron population in the metal surface, and the third is due to the effect of the adsorbate on the free-electron

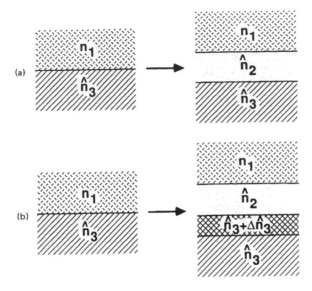

FIG. 31.   Schematic models considered in the analysis of optical effects generated by adsorbed layers or by very thin films.   (a) Simplest single-film model.   (b) Dual-film model that considers the perturbation of the properties of the substrate due to the adsorption and, in electrochemical applications, the modulation of the electrode potential.

collision time in the metal surface.   Figure 31 describes a model used by various workers for the analysis of optical effects due to chemisorption which contains an additional effective film generated due to the modulation of the optical properties in the metal surface.   However, such models are essentially phenomenological, and equations such as Eqs. (32a) and (32b) cannot be used with full confidence for the resolution of the elusive optical properties of the chemisorbed oxide itself.

A somewhat more realistic approach to the description of the optical properties of an interfacial region was given by Plieth [84,85].   He introduced, by analogy with the thermodynamic treatment of surface regions, surface-excess functions for the continuously varying optical properties in the direction z normal to the surface.   This approach was suggested specifically for the description of the electrode-electrolyte interface. Plieth also stressed the uniaxial anisotropy expected and thus chose for complete optical description the real and imaginary parts of the following integrals:

$$\text{(I)} \int (\varepsilon_t - \varepsilon_{ref}) \, dz \qquad \text{(II)} \int \left(\frac{1}{\varepsilon_n} - \frac{1}{\varepsilon_{ref}} \right) dz \qquad (33)$$

where ε_t and ε_n are the tangential and normal components of the dielectric tensor, and where the integration is carried out from the homogeneous ambient phase to the homogeneous substrate phase, and where ε_{ref} represents the reference state of the film-free substrate. According to Plieth [84,85], the simplest model of a single film, as described above for the case of oxygen chemisorption, amounts to a definition of an effective dielectric function for an interfacial region with an overall effective thickness d_1, by

$$\langle \varepsilon \rangle = \frac{1}{d_1} \int \delta \varepsilon(z) \, dz \qquad (34)$$

and equations such as Eqs. (32a) and (32b) assume a model of two stratified films with two different effective dielectric constants and thicknesses.

The "integral model" of Plieth [84,85] sheds some light on the meaning of effective dielectric constants and effective film thicknesses evaluated from macroscopic single-film or dual-film approaches to ellipsometry of chemisorbed layers. It does not, however, treat the problem of the relationship between the macroscopic quantities $\hat{n}_f$, d_f, and the microscopic parameters of the adsorbate, such as the molecular polarizability, diameter, cross-sectional area, or degree of coverage by the adsorbed atoms or molecules. Some microscopic theories, which describe the dielectric tensors of adsorbed layers in terms of molecular properties and coverages, have been developed by several authors [86—89]. However, in most cases an essentially macroscopic model has been employed in the analysis of ellipsometric results in the submonolayer domain. This was done by extrapolation of the Lorentz-Lorenz equation, derived

originally for a model of a cavity in a continuous dielectric, for the
description of a submonomolecular film. This approach yields the
following expression [87]:

$$\frac{n_\theta^2 - 1}{n_\theta^2 + 2} = \frac{4}{3\pi} \sum_i \frac{\alpha_i \theta_i}{\sigma_i d_i} \tag{35}$$

where θ_i is the surface coverage, α_i the molecular polarizability, σ_i the
molecular cross-sectional area, and d_i the molecular diameter of species
i. This treatment, as well as all the other theoretical approaches
attempted based on adsorbate molecular properties [81], are strictly
justified only when there is virtually no electronic interaction between
substrate and adsorbate (e.g., physisorption). These approaches are
probably reasonable in electrochemical systems for the description of the
electrolyte side of the electrode-electrolyte interface in the presence of
electrostatic interactions alone, such as nonspecific ionic adsorption.

Let us return to some applications of ellipsometry for measuring
adsorbed layers. A noticeable interest in oriented monolayers on metal
and nonmetal surfaces has developed recently, particularly around the
unique features of Langmuir-Blodgett films. Some interesting ellipsometric work has been performed in this field. The work of Smith [89]
has demonstrated that ellipsometric measurements of $\delta\Delta$ for submonolayer
films of various organic compounds on mercury surfaces gave good linear
relation with surface coverage, as derived independently from measurements of surface tension in a Langmuir trough. Phase transitions in a
two-dimensional monomolecular layer induced by high surface pressures
could also be detected ellipsometrically, and the molecular orientation was
shown to result in optical anisotropy in the film [89]. The theory for
[90,91] and ellipsometric investigations of [92,93] anisotropic monolayer
films made of oriented large organic molecules has been described in
detail. Langmuir troughs were used to vary the surface coverage and
orientation of such adsorbates. In the cases mentioned above, the

quasi-macroscopic Lorentz-Lorenz treatments were employed to calculate
the two orthogonal refractive indices in the unisotropic film, based on
independently acquired data of molecular dimensions and polarizabilities
[92]. (For more recent references on the use of ellipsometry for
studying LB films, see Refs. 144 to 146.)

The use of the variation in Δ as a "coverage gauge" in electrochemical systems (see Sec. IV.B) has been described in detail for
bromide adsorption on platinum [25]. The general conclusion was that
the Δ parameter is a sensitive probe of anionic adsorption, particularly
when the coverage changes sharply with the applied potential. The
$\delta\Delta$ vs. V plots recorded during scans of the potential in the double-
layer and Pt-H regions for a Pt electrode immersed in solutions containing bromide have the qualitative form of the bromide adsorption isotherm.
However, the difficulty with quantitative evaluatuion of the anionic
coverage is that the polarizability of the anion has to be known. This
parameter is unknown for strongly interacting ("chemisorbed") ions, the
adsorption of which is associated with partial charge transfer and with
possible effects of place exchange that may take place at higher concentrations and anodic potentials [94]. These results demonstrate that
for strongly interacting chemisorbed species (which are the rule rather
than the exception in the case of solid-electrolyte interfaces), ellipsometric measurements on monolayers or submonolayers can be quite
informative, although difficult to quantify in terms of surface coverages
(see the next section).

2. *Ellipsometric Measurements of the Metal-Electrolyte Interfaces*
    ("Double-Layer Effects")

Processes at the metal-electrolyte interface, particularly ionic and
molecular adsorption on a submonomolecular level, were at the center of
the activity in combined ellipsometric-electrochemical measurements performed in the late sixties and early seventies. Since then the activity
has shifted toward thicker surface films. One reason is the fast develop-

ment of in situ Raman [95] and infrared [96] techniques for the examination of electrode surfaces. These techniques proved to be very valuable for the in situ study of the adsorption of organic molecules on electrode surfaces. The other reason is that the interpretation of ellipsometric results obtained for surface processes at metal electrodes on the submonolayer level is quite complex, as explained in Sec. V.B.1. Consider, for example, an experiment in which the potential of a Pt electrode immersed in an acid aqueous solution is switched from a cathodic bias to an anodic bias, so as to induce and study the adsorption of anions. At least two additional effects of potential modulation on the optical properties of the metal-electrolyte interface have to be considered when the results are evaluated. These effects are the modulation of the components of the dielectric constant of the metal substrate, due to the surface charging and due to band bending in a thin surface layer of the metal electrode (ca. 0.2 nm), and the modulation of the refractive index of the water layer in contact with the electrode surface, due to possible changes in water density caused by the effect of water dipole reorientation [25,97,98]. Each of the foregoing perturbations in the optical properties of the interphasial domain can be described in terms of a film with an effective thickness and an effective refractive index. Each of the individual optical effects considered is calculated to bring about variations in the ellipsometric parameters of the order of $10^{-2}-10^{-1}$ degree per 1 V of change in applied potential [25]. The magnitude of these ellipsometric effects is small because of the relatively small perturbations in optical parameters taking place in films with effective thicknesses of 0.1−0.2 nm. However, with automated ellipsometers, signal/noise enhancement can be effectively achieved in measurements of such small effects in the double-layer region, by averaging the readings for many (e.g., 100) repetitive triangular potential cycles or pulses [25]. Such repetitive cycling also helps to maintain an electrode surface that is free of impurities. Having recently used such an experimental approach,

Hyde et al. reported the following results [25]. For a Pt electrode in
an aqueous electrolyte with minimal specific anionic adsorption, e.g.,
dilute perchloric acid, the positive sign and the angular dependence of
the variation of Ψ with increased positive potential within the double-
layer region were consistent with the formation of a transparent layer
on the electrolyte side of the interface with a real n_f larger than that of
the electrolyte (see Fig. 32 and the discussion in Sec. IV.B). From calculation of the maximum optical effects expected for perchlorate adsorption,
found to be much smaller than the measured ellipsometric effects, it was
concluded that the variation of Ψ with V within the double-layer region is
due to a "water densification" effect described above, which corresponds
to an increase of the surface density of H_2O molecules by about 10% for
an increase in the electrode potential between 0.4 and 0.8 V. It is
interesting to compare these ellipsometric results for the Pt-H_2O interface with those of measurements of surface plasmon excitation at the
metal-electrolyte interface in the double-layer region, as described by
Tadjeddine [99]. This author described such measurements for gold and
silver electrodes, both "free-electron-like metals" as required in measurements based on surface plasmon excitation. Tadjeddine's conclusion was
that effects of the reorientation (densification) of the H_2O layer were
negligible at gold and silver electrodes. This conclusion was apparently
based on the detection of an appreciable shift in the real part of the
wave vector only on the positive side of the point of zero charge, thus
ascribed fully to anionic adsorption [99]. However, as pointed out by
Hyde et al. [25], it is quite likely that the optical effect due to "H_2O
densification" in the double-layer region would also be strongly asym-
metric with respect to the potential of zero charge (pzc). Furthermore,
in the case of Pt in dilute perchloric acid solutions, the optical effects
in the double-layer region, all at positive potentials vs. the pzc, could
not be accounted for by anionic adsorption alone.

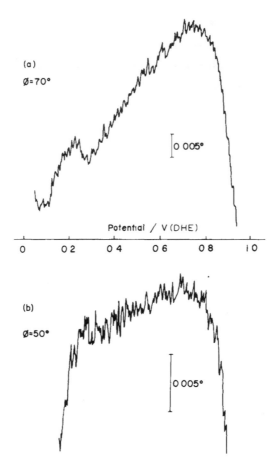

FIG. 32. Ψ vs. V curves at 400 nm obtained for a Pt electrode in 0.1
M $HClO_4$, shown on an amplified scale, for the double-layer region and
the onsets of the hydrogen and oxygen regions. The scan at 50 mV
sec^{-1} was applied between 0.09 and 1.30 V vs. DHE. Scale bar: 5 ×
10^{-3} degree in Ψ. 100 averaged scans. (a) ϕ = 70°; (b) ϕ = 50°.
These results demonstrate the positive $\delta\Psi/\delta V$ in the double-layer region,
which is measured at high angles of incidence, but goes to zero at ϕ_1
close to 45°. (From Ref. 25. Reprinted with permission from Elsevier
Science Publishers, Amsterdam.)

For aqueous solutions containing bromide anions, it was demonstrated
by Hyde et al. [25], that by employing signal averaging, very reproducible and noise-free Δ vs. V curves could be obtained which reflected
directly the variation of θ_{Br^-} with applied potential. The variation of
anionic coverage with V could be thus followed in situ, throughout both
the double-layer and Pt-H potential regions. The results of Hyde et al.
are shown in Fig. 33. In this case of ellipsometric investigation in the

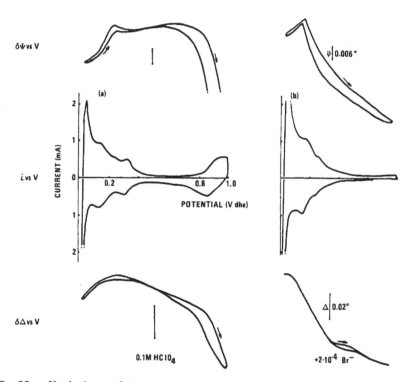

FIG. 33. Variations of the current and the ellipsometric parameters Δ
and Ψ with potential, during a triangular scan in 0.1 M HClO$_4$, before
and after adding 2×10^{-4} M Br$^-$. In the presence of Br$^-$, the $\delta\Delta$ vs. V
curve directly tracks the variation of θ_{Br^-} with V within the Pt-H region
(more negative Δ readings correspond to higher θ_{Br^-}). Also, $\delta\Psi/\delta V$
becomes negative in the double-layer region due to the bromide chemi-
sorption process. (From Ref. 25. Reprinted with permission from
Elsevier Science Publishers, Amsterdam.)

double-layer region, the enhancement of signal/noise by repetitive
multicycling and by computer-aided averaging allowed a dramatic improvement in the quality of the results compared with earlier steady-state
ellipsometric measurements of anionic adsorption [100,101]. Thus quite
different conclusions regarding the adsorption isotherm were arrived at.
The dependence of θ_{Br^-} on V in acid solutions, as reported by Hyde
et al., is in very good agreement with recent ex situ measurements by
Hubbard and co-workers [142].

Hyde and Gottesfeld described later time-resolved ellipsometric
measurements performed during potential-driven anionic adsorption-
desorption pulses [94]. They showed that gradual steps in the overall
process of bromide adsorption on Pt can be detected in time-resolved
ellipsometric measurements. While in 10^{-4} M solutions the electrosorption
of bromide was shown to be diffusion controlled, in 10^{-3} M bromide
solutions the anodic electrosorption was shown to proceed in two dis-
cernible steps. The first caused a decrease of Δ and the second a
subsequent increase of Δ, as shown in Fig. 34. These two steps were
interpreted as anionic adsorption followed by reconstruction of the
surface. It was suggested that the reconstruction apparently involves
a partial transformation from an overlayer of adsorbed bromide to a
mixed layer of Br and Pt atoms. There is a parallel to be found in the
adsorption of Br_2 from the gas phase onto a Pt surface, which is also
reported to induce a place exchange at the submonolayer coverage
level [102].

C.   Optical Properties and the Analysis of Composite Surface Films

Solution-filled voids, grain boundaries, disordered regions, and other
inhomogeneities on the microstructural length scale of 1–100 nm significantly affect the optical properties of films in the visible-near UV region.
This is true even when the films are macroscopically uniform. The
interpretation of ellipsometric results for such "imperfect" films was

limited in the past to semiquantitive treatments and was considered an impediment in ellipsometric work on more complex interfaces. Recent theoretical and experimental work, e.g., by Hunderi [103] and particularly by Aspnes and co-workers [104], has allowed detailed interpretation of ellipsometric results for composite surface films. In fact, spectro-ellipsometry has now become a quantitative source of information on the microstructure within composite surface layers obtained in a nondestructive in situ measurement [104].

The simplest case of a surface film with internal microstructure considered in quantitative treatments is that of a two-component composite. For this case, the following general expression can be written for the dielectric function (see, e.g., Ref. 104):

$$\frac{\varepsilon - \varepsilon_H}{\varepsilon + \kappa\varepsilon_H} = f_a \frac{\varepsilon_a - \varepsilon_H}{\varepsilon_a + \kappa\varepsilon_H} + f_b \frac{\varepsilon_b - \varepsilon_H}{\varepsilon_b + \kappa\varepsilon_H} \tag{36}$$

where f_a and f_b are the volume fractions of components a and b in the composite ($f_a + f_b = 1$), ε_a and ε_b the dielectric functions of phases a and b, and ε_H the dielectric function of the "host phase." Specific relevant structures to be considered here are composite layers with one solid phase (e.g., metal, oxide, or polymer) and one ambient (electrolyte) phase, where the first will be designated by the subscript a and the second by subscript b. For a cermet microstructure, where solid particles are coated by the electrolyte phase, $\varepsilon_H = \varepsilon_b$. For the aggregate microstructure consisting of a random mixture of solid and ambient "particles," $\varepsilon_H = \varepsilon$, and for the solid skeleton (Swiss-cheese like) microstructure, $\varepsilon_H = \varepsilon_a$. The first and third cases correspond to Maxwell-Garnett effective-medium models, and the second is the Bruggeman effective medium approximation. The screening parameter κ depends on the dimensionality of the elementary "particles" in the microstructure. Typically, κ assumes a value close to 2, corresponding to three-dimensional "particles." The other special case of Eq. (36) is the Lorentz-Lorenz effective-medium expression, which corresponds to the mixing of

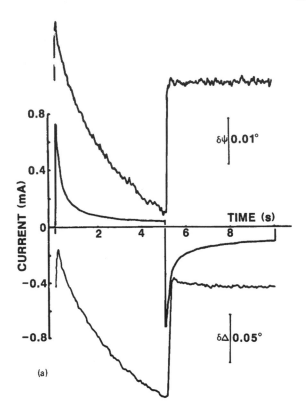

FIG. 34. (a) Variations of Δ and Ψ with time following a potential step-ping experiment from 0.0 to 0.7 V and back, in 0.1 M $HClO_4$ + 2.10^{-4} M Br^-. Diffusion-controlled adsorption and "instantaneous" desorption are observed. (b) A peculiar behavior of $\delta\Delta$ vs. time following a step from 0 to 0.7 V, found in more concentrated bromide solutions (2×10^{-3} M): Δ first drops due to the increase of θ_{Br^-}, but this is followed by a slower positive shift of Δ. Interpretation suggested: slow field-assisted incorporation of adsorbed bromide into the outermost layer of Pt atoms. (From Ref. 94. Reprinted with permission from North-Holland Physics Publishing, Amsterdam.)

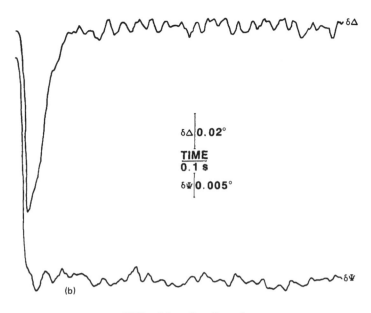

FIG. 34.   Continued

of a and b on an atomic scale, as opposed to the mixing of regions
large enough to possess their own dielectric identity.   The Lorentz-
Lorenz expression is based on a model of a lattice of points that are
randomly assigned two different polarizabilities, α_a and α_b, related to
dielectric functions ε_a and ε_b of the phases a and b in their pure forms
according to the Clausius-Mossoti equation.   It is also assumed in the
Lorentz-Lorenz expression that vacuum ($\varepsilon = 1$) is the host medium, and
Eq. (36) thus takes the form (for $\kappa = 2$)

$$\frac{\varepsilon - 1}{\varepsilon + 2} = f_a \frac{\varepsilon_a - 1}{\varepsilon_a + 2} + f_b \frac{\varepsilon_b - 1}{\varepsilon_b + 2} \qquad (36a)$$

Equation (36a) and its extrapolation to N components are employed for
the analysis of optical effects of electrosorbed layers in electrochemical
systems [see Eq. (35)].

Examination of Eq. (36) shows that the optical properties of com-
posite films with given ε_a and ε_b may depend in a dramatic way on the

volume fractions of the phases and the interconnectedness of the micro-
regions of a given phase, such as a random mixture or a Swiss-cheese-
like microstructure. An interesting example has been described and
analyzed ellipsometrically by Heller et al. [105], for the case of Pt films
photoelectrochemuically deposited on InP substrates. In the latter case,
deposition of Pt films as thick as 20−30 nm which are almost completely
transparent in the visible, has been acomplished by employing unstirred
dilute solutions of $Pt(OH)_6^{-2}$(aq) in the photoelectrochemical deposition
process. The significance of this result in the context of the efficiency
of photoelectrochemical H_2 generation is that Pt films on InP serve as
efficient electrocatalysts, yet transmit 92% of the light at wavelengths
between 210 and 750 nm, as required for excitation of electron-hole
pairs within the surface of the InP phase. Spectroscopic ellipsometry
has been used as the principal tool for the characterization of such
composite Pt-electrolyte films, supplying information on the volume
fraction and the microstructure of Pt in the deposited layers. The
conditions and procedure for obtaining such microstructural information
from ellipsometry can be summarized as follows [104]: The first condition
is that accurate spectra of ε, over a reasonably wide energy range,
should be known for the surface film and for its constituent phases.
In the specific case considered, $\varepsilon_f(\lambda)$ (and d) for the composite Pt film
was evaluated from spectroellipsometric measurements performed on the
Pt/InP system [see Fig. 35(a)]. The ε spectrum of bulk dense Pt was
also recorded ellipsometrically [105] and is shown in the same figure.
The second condition is that the individual dielectric responses of the
constituent phases (Pt and electrolyte in the specific case considered)
can be combined through one of the effective-medium models described
above to reproduce the measured dielectric response of the composite,
employing only the two wavelength-independent parameters that appear
in Eq. (36), i.e., the volume fraction and the screening parameter, as
well as a wavelength-independent thickness. In the case considered

[105], the analysis of a 250-point spectrum between 1.5 and 5.5 eV showed that the data for the composite film were consistent with the aggregate structure (random mixture of metal plus ambient "particles"). The data were thus best described by the Bruggeman approximation. The best fit of the dielectric dispersion shown in Fig. 35 to the Bruggeman model gave d_f = 32.6 ± 0.8 nm, κ = 0.80 ± 0.08, and f_a = 0.50 ± 0.02 [105]. Figure 35(a) demonstrates the change in sign of ε_1 from negative, as typical for a metal (see Sec. IV.A), to positive. It also shows the general strong decrease in the magnitude of the dielectric constant when comparing dense Pt with the composite Pt-electrolyte layer. This is the result of the severe loss of metallic characteristics in this Pt film. Figure 35(b) and (c) are SEM and TEM micrographs for the same Pt film deposited on InP [105]. The results of the (destructive) microscopic examination confirmed the conclusions on the microstructure derived with the (nondestructive) spectroellipsometric tool [105].

Another example of the use of spectroscopic ellipsometry and of Eq. (36) in the evaluation of optical properties of composite surface layers has been reported by Gottesfeld et al. [106] for the process of roughening a metal alloy surface by anodic dissolution. The authors found that a metal-deficient surface layer can form under some electrochemical conditions on Cr-rich PtCr alloy surfaces and can reach maximum thicknesses of tens of nanometers. The dielectric spectrum of such a metal-deficient layer could be evaluated ellipsometrically to yield information on the volume fraction of the metal skeleton and on the degree of interconnectedness of the metal domains. The results are shown in Fig. 36, which demonstrates ε spectra for bulk Pt and Cr metals, for the bulk $Pt_{0.2}Cr_{0.8}$ alloy, and for the metal-deficient surface layer formed on the latter alloy. The ε spectrum for the surface layer shown in Fig. 36 looks like that of a "diluted metal." In contrast to the case presented in Fig. 35, the value of ε for the surface layer, although significantly smaller in magnitude than that of the bulk, is still negative.

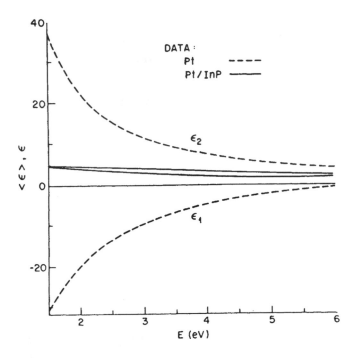

FIG. 35. (a) Solid curve, ε spectrum of the Pt film photoelectrochemically deposited on InP, calculated by solving a single-film model using
the results of spectroellipsometry with d_f = 33.7 nm. Dashed curve,
reference ε spectrum of dense Pt film prepared by electron beam evapora-
tion. (b) Scanning electron micrograph of the transparent Pt film with
the dielectric response shown in (a). (c) Transmission electron micrograph of the same Pt film. (From Ref. 105. Reprinted with permission
from the Journal of Physical Chemistry, copyright 1985, The American
Chemical Society.)

The analysis, according to Eq. (36), showed in this last case a much
better fit to the Maxwell-Garnett version, with the metal as the host
phase ("Swiss-cheese" microstructure), which implies good connectedness
between the metallic domains. Thus, while the volume fraction of the
metallic particles was only 30% in the last case [106], the better connectedness between the small metal domains resulted in stronger metallic
properties than those of the film described in Fig. 35, where the fraction of the metal was higher (50%) but the degree of connectedness was
poor.

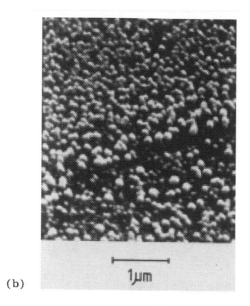

(b)

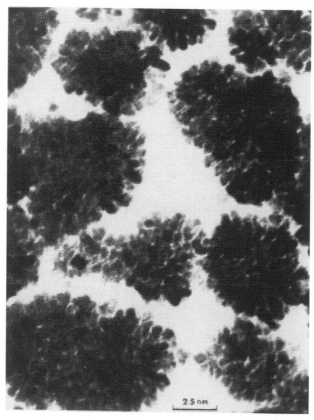

(c)

FIG. 35. Continued

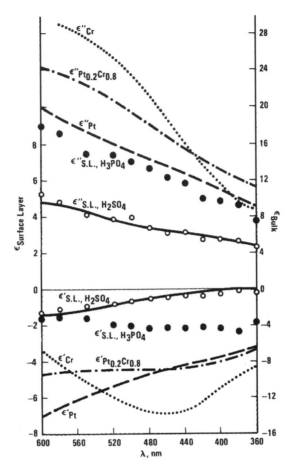

FIG. 36. Optical dielectric constant spectra of the surface layers (S.L.) which develop on the $Pt_{0.2}Cr_{0.8}$ alloy due to cycling in dilute H_2SO_4 and in 85% H_3PO_4, as evaluated ellipsometrically, presented together with the corresponding spectra for bulk Pt, Cr, and $Pt_{0.2}Cr_{0.8}$. Treatment of the surface layer spectrum in terms of effective-medium theory showed that the composite surface layer can be described in terms of a Maxwell-Garnett model with the metal (host) phase occupying 30% of the volume and the metal domains well interconnected. (From Ref. 106. Reprinted with permission from Elsevier Science Publishers, Amsterdam.)

The composite nature of surface films in electrochemical systems is not confined to metal-electrolyte microstructures. For example, Kang and Shay have analyzed the optical spectrum of Ir oxide films produced by reactive sputtering of iridium in argon-oxygen mixtures [107]. Such oxide films have a volume fraction of 0.28 occupied by IrO_2 [107]. The authors deduced that the absorption peak observed at 610 nm for such blue films is due to a resonance absorption expected at

$$\omega_o = \omega_p \left(1 + \frac{2 + f}{1 - f} \, \varepsilon_2 \right)^{-1/2}$$

where ω_p is the plasma frequency for metallic single-crystal IrO_2, f the volume fraction of IrO_2 in the composite film, and ε_2 the dielectric constant of the aqueous (host) phase in a Maxwell-Garnett composite. Recent ellipsometric examination of films of polyaniline has revealed optical properties somewhat similar to those of hydrous Ir oxide films [76,79]. Particularly, the peak at about 700 nm that appears in the colored (anodic) form of such films strongly resembles in location, magnitude, and shape the absorption peak for blue sputtered Ir oxide. In the absence of parameters such as the plasma frequency for dense (metallic?) polyaniline, it can be argued only that the similarity of the spectra of the colored forms of polyaniline and of the hydrous oxide, plus the similarity in the level of k_f, suggest the possibility of an analogous origin, such as a Maxwell-Garnett composite, for the optical behavior of polyaniline films grown in aqueous solutions.

D.   Metal Electrodeposition

1.   UPD Metal Layers

Optical characterization of the interface during the electrodeposition of metallic layers on metallic substrates includes examination of mono- and submonolayer coverages (i.e., UPD layers) and of thicker electrodeposits reaching 10–100 nm. Interpretation of ellipsometric results for sub-

monomolecular coverages is not straightforward (see Sec. V.B). However, in some of the cases of the metallic monolayers investigated, the
apparent optical properties obtained by using a single-film model better
resemble the optical properties of the corresponding bulk materials.
For example, Farmer and Muller argued recently that for the case of
UPD Pb on Ag(111), the variations of Δ and of Ψ with coverage can be
interpreted with an "island model" based on a single effective film [108].
In their model, two-dimensional clusters of metal adatoms were treated
as islands of constant thickness and complex refractive index, which
grow and coalesce to form the monoatomic layer. This model predicted
the measured linear relation between $\delta\Delta$, or $\delta\Psi$, and the coverage, θ, in
the UPD domain, where θ was derived from the cyclic voltammogram.
Using this model, Farmer and Muller found for UPD Pb on Ag(111) at
λ = 514.5 nm, $\hat{n}_{Pb}$ = 1.28-4.08i, and for UPD Pb on Cu(III), $\hat{n}_{Pb}$ =
1.22-3.52i. The apparent film thicknesses at θ = 1 were 5.15 ± 0.03
and 4.03 ± 0.02 Å, respectively [108]. Compared to the case of oxide
films (Sec. V.B), the apparent optical properties change less between
the monolayer level and the thicker deposit of Pb. A typical variation
is from 0.95-3.99i for the UPD layer at θ = 1 to 1.99-4.28i for a
layer of bulk Pb 19.4 nm thick, both grown on the same Cu substrate.
Some of this difference was ascribed to the microporous nature of the
thicker Pb deposit [108]. The choice of the "island model" was made
for the Pb/UPD layer [108] because a Bruggeman version of effective
medium theory [Eq. (36)] could not predict the measured linear depen-
dences of $\delta\Psi$ or $\delta\Delta$ on θ. (The more relevant version of effective-
medium theory for the case of mixing on an atomic scale [104], as suggest-
ed by the Langmuirian Pb isotherm [108], is the Lorentz-Lorenz equation [Eq. (36)].) This last example of a UPB metal layer is remarkable
in that a model based on a single film, with optical properties determined
by weighted contributions of the dielectric constants of the metal deposit
and of the electrolyte, could reasonably well describe the optical behavior

of the submonomolecular layer.   In principle, interaction of the adsorbed
UPD atoms with the substrate should bring about several additional
complicating optical effects, as described in Sec. V.B.   The apparent
explanation is that in this case these additional effects are relatively
small.   However, this is definitely not a general rule for UPD metal
layers.   A counterexample is the spectroellipsometric work of Chao and
Costa done on UPD Ag layers on gold [109].   These authors attempted
to correct carefully for substrate surface effects, including the effect of
applied potential on the free-electron population and on the free-electron
collision time in the metal substrate surface, and the effects of the field
on interband transitions in the gold surface.   Following such corrections,
the spectrum of the silver layer itself was evaluated for the UPD Ag
layer.   The spectral features were very different from those of bulk
Ag and were interpreted in terms of both Ag-Ag and Ag-Au interactions.
These results are described in Fig. 37.   Apparently, the necessary
condition for the dominance of the optical properties of the UPD metal
itself is that the metal deposited has a pronounced featured spectrum in
the λ domain probed, and the substrate is associated with a featureless
spectrum.   For example, for UPD Cu on Pt, the onset of Cu-Cu bonding
about $\theta = 0.56$ already brings about characteristic features in the ε''
spectrum of the deposited layer, peaking at about 560 and 620 nm [110].

   For completeness, the optical work done in the past on UPD metal
layers employing reflectance measurements, rather than ellipsometry,
should also be described briefly.   The reasons for confining such optical investigations in the past to reflectometric measurements were the
simpler instrumentation required to measure only light-intensity variations in a wide λ domain, and the unavailability of automated wavelength-
scanning ellipsometers.   If reflectance measurements are taken in a
sufficiently wide λ range, Kramers-Kronig analysis can yield the phase
parameter (see e.g., Ref. 111), while this parameter (i.e., Δ) is directly
measured at each wavelength by ellipsometry.   Reflectance work on UPD

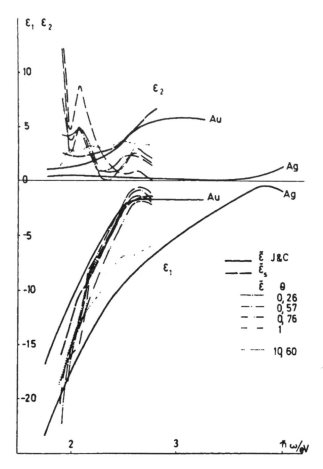

FIG. 37. The ellipsometrically evaluated energy dependence of the dielectric constant of silver films deposited on gold, as a function of silver coverage. For comparison the dielectric constant of the effective gold substrate surface at 1.2 V (ε_s), as well as ε_{Au} and ε_{Ag}, are also shown. (From Ref. 109. Reprinted with permission from North-Holland Physics Publishing, Amsterdam.)

metal layers is summarized in a review article by Kolb [111] in which
there is an example of a clear change in the apparent optical properties
between the UPD Ag layer and the thicker Ag layer deposited on Pt
[112]. (A single film is assumed in the analysis.) In λ regions close
to pronounced features in the spectrum of the substrate, strong effects
due to modulation of the substrate properties during deposition of the
UPD layer have been found. Such substrate modulation effects result
in shapes of $\delta R/R$ spectra, which are derived from the spectrum of the
substrate. Examples are UPD layers of Pb, Tl, and Cu on Au, all of
which generate $\delta R/R$ spectra that when analyzed under a single-film
assumption, exhibit characteristic spectra of a distorted gold surface
[113]. On the other hand, deposition of submonomolecular layers of
metals with pronounced spectral features onto substrates with a feature-
less ε spectrum in the same domain (e.g., Cu/Pt) yields ε spectra with
features characteristic of the "layer itself." The anisotropy of UPD
layers has been stressed by Dignam and Moskovitz, who have derived
quantitative expressions using linear approximations for relative changes
in the reflectance arising from the deposition of a uniaxial film onto an
isotropic substrate [114]. As an interesting consequence of anisotropy,
it has been pointed out that a thin, transparent, but anisotropic film on
a metallic substrate yields $\delta R/R$ values that, when calculated assuming
an isotropic film, may pretend absorbing properties [115]. It has also
been shown for UPD Pb on gold that direct proportionality between $\delta R/R$
and the coverage θ may be obtained, even in cases in which the apparent
optical properties result mostly from a substrate modulation effect [116].
Such behavior can arise if, for example, the optical effect due to perturbation of the substrate surface properties varies linearly with charge,
and thus scales with the coverage by the metal atoms deposited. In
any case, these results show that proportionality between the variation
of the measured optical parameter ($\delta R/R$, $\delta\Delta$, or $\delta\Psi$) and the coverage,
θ, in the submonolayer region is not a guarantee that the optical effect

is due to the "film itself." Although the last examples for optical work on UPD metal layers are taken from reflectometric investigations, the conclusions with regard to the interpretation also apply to ellipsometric investigations of submonomolecular layers of metal deposits.

2. Thick Metal Deposits

Recent ellipsometric and spectroellipsometric investigations of thicker metal deposits (10—100 nm) have been devoted to gradual development of the optical properties of deposits during electrochemical growth. From variations of the optical properties with increase in amount of metal deposited, conclusions about the morphology of the deposit and the mechanism of its growth have been derived. An interesting attempt has been described by Abyaneh et al. [117] to study the initial steps of the electrocrystallization of nickel by simultaneous ellipsometric and amperometric investigation during potentiostatic deposition. Their results are shown in Fig. 38. Theoretical equations were derived for Δ and Ψ transients assuming two possible growth forms based on conical or on hemispherical nucleation centers. The better fit was obtained for the model that assumed growth from hemispherical centers. The model used by Abyaneh et al. to describe a deposit with a given volume fraction of metal was the Lorentz-Lorenz version of effective-medium theory, but other versions may be more applicable. In fact, the best description of the deposit may vary during its development. In the earliest stages, when the hemispherical islands have still not connected, the Bruggeman model may apply, but for somewhat thicker deposits the Maxwell-Garnett version may become applicable (Sec. V.C). Despite these limitations, the time-resolved ellipsometric monitoring of the developing metal deposit seems to supply valuable information about the topography of the growth centers. Farmer and Muller have recently described a spectroellipsometric investigation of Pb deposits up to 88 nm thick on copper substrates [118]. The thicker Pb deposit develops on top of the UPD layer, the optical characterization of which is described in Sec. V.D.1.

The simulated spectra of Δ and Ψ for a layer of bulk Pb on copper with a thickness calculated according to the charge consumed in the deposition did not agree at all with the measured Δ and Ψ spectra for deposited Pb layer thicknesses below 110 nm. For such thicknesses, the Δ and Ψ spectra could be simulated by employing the Bruggeman model (Sec. V.C), where the volume fraction of Pb found in the deposited layers ranged between 0.35 and 0.56, as shown in Fig. 39. Farmer and Muller also showed that the improved quality of the Pb deposit, achieved by using rhodamine B as a plating additive, could be demonstrated from the spectroellipsometric measurements: The Pb films were found from the analysis to reach a metal volume fraction of 0.90, following the addition of rhodamine-B. The differences between the ellipsometric effects due to the growth of Pb films in the presence and in the absence of the plating additive are quite striking, as shown in Fig. 40. It is immediately noticeable that the porous deposited metal layer, which has lost metallic properties due to loss of good connectedness between the metal particles, generate negative changes of Δ that are typical for the growth of dielectric layers (see, e.g., Fig. 12). On the other hand, the much better metallic properties of the Pb layer obtained in the presence of the plating additive bring about positive variations in Δ.

The strong deviation from bulk metal properties found for the films of thicknesses below 100 nm is not unique for electrodeposited layers. Aspnes et al. have investigated sputtered gold films with spectroellipsometry and revealed their microporous nature [52]. More recently, Aspnes and Craighead have described a thorough spectroellipsometric investigation of evaporated Rh films of two thicknesses, 15 and 30 nm, prepared at a background pressure in the mid 10^{-6} torr range [41]. The dielectric spectrum of the Rh films derived from such ellipsometric investigations could be explained only with a cermet model, in which metal particles are isolated by an insulating oxide coating. This model was later verified by (destructive) investigation with a transmission

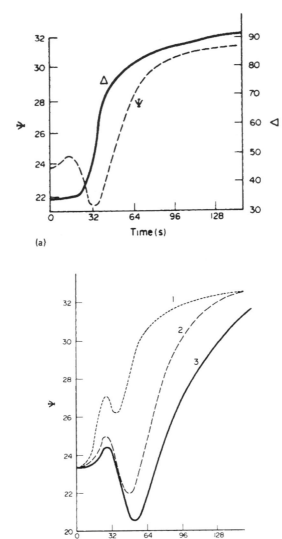

(a)

(b)

FIG. 38. (a) Measured Δ and ψ transients due to the electrodeposition of nickel on a vitereous carbon substrate during a potential step from E = −120 mV to E = 900 mV (vs. SCE). Solution composition 0.1 M Ni₂SO₄, 0.6 M NaCl, 0.58 M H₃BO₃. (b) and (c) Calculated ψ and Δ transients, respectively, for a hemispherical growth model and growth rate of 9 Å sec⁻¹ (curve 1), 3 Å sec⁻¹ (curve 2), and 2 Å sec⁻¹ (curve 3). (From Ref. 117. Reprinted with permission from *Electrochemica Acta*, copyright 1983, Pergamon Journals Ltd.)

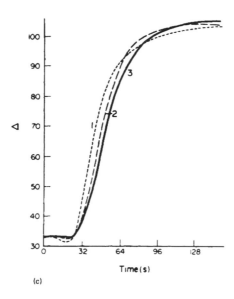

(c)

FIG. 38.   Continued.

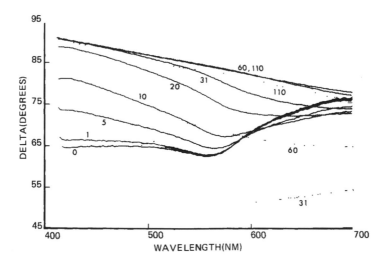

FIG. 39.   Solid lines, simulations of Δ spectra for compact Pb single films
growing on a copper substrate.   Simulated spectra for Pb deposits of
thicknesses 0, 1, 5, 10, 20, 31, 60, 110 nm are shown.   Dotted lines,
experimental measurements for 31, 60, and 110 nm equivalent deposit
thicknesses.   (From Ref. 118.   Reprinted by permission of the publisher,
The Electrochemical Society, Inc.)

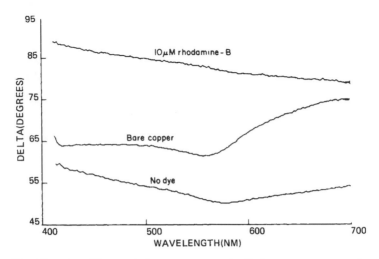

FIG. 40. Spectroellipsometric measurement of the parameter Δ for a Cu substrate covered by thin Pb deposits corresponding to a 31-nm-thick compact Pb film. The effect of a plating additive on the spectrum is demonstrated. The bare Cu substrate spectrum is also shown. (From Ref. 118. Reprinted by permission of the publisher, The Electrochemical Society, Inc.)

electronmicroscope, which demonstrated the small (5 nm), isolated metal particles in the evaporated Rh film. This work by Aspnes and Craighead [41] gives a very detailed account of a spectroellipsometric analysis of a composite film, and it is a good source of further information about the accuracy obtainable in the determination of optical properties and thicknesses with state-of-the-art instrumentation. Two further examples for the analysis of composite metal-electrolyte films, one generated by photoelectrochemical deposition of Pt onto InP [105] and another by the dealloying of a Pt-Cr alloy surface [106], have been described in Sec. V.C.

E. Ellipsometric Investigations of Corrosion and Passivation Processes

The interfacial processes taking place during metal corrosion create the most complex situations investigated in electrochemical systems by means of ellipsometry. These processes usually involve simultaneous events of metal surface roughening, film formation, and accumulation of solution products due to metal dissolution processes [119]. All of these electrochemical (or chemical) processes bring about substantial ellipsometric effects. Muller and Smith described the complicated optical system that arises during the anodic oxidation of silver in alkaline solutions. The interfacial processes of metal roughening, formation of a solution boundary layer through which dissolution products diffuse away from the surface, growth of three types of surface films with increased states of hydration in the direction away from the metal surface, and growth of secondary crystals due to nucleation from solution had to be included in a model that accounted for the ellipsometric results. Although some of these events could be resolved in time, which helped somewhat in the analysis, the example obviously clarifies how complex the situation can become in corroding metal systems. Similar cases of complex interfaces, in which several films exist one on top of the other, have been analyzed recently by spectroellipsometry in the case of semiconductor substrates. For example, following laser annealing of Si-H layers, such analysis considered three or four surface films with various volume fractions of Si-H, crystalline Si, SiO_2, and voids [120]. However, in the case of a corroding metal in an aqueous solution, the assignment of values of n to the components of a composite film is usually less certain, because of unknown degrees of hydration and/or nonstoichiometry.

The other extreme situation in the ellipsometric research of corrosion and passivity is one in which electrochemical conditions can be found, such that the growth of a passive film takes place at a 100% charge efficiency, without any accompanying dissolution, precipitation, or substrate roughening processes. Such electrochemical conditions can be defined, for example, for the case of the potentiostatic growth of the passive film on iron. Chen and Cahan [18] examined the passivation film on iron in borate buffer, using in situ spectroscopic ellipsometry and reflectometry to obtain spectra, thickness, and kinetics for the film formation and for film reduction processes under various electrochemical conditions. Although the thickness of the passive film on iron does not exceed several nanometers, the authors found from the ellipsometric data that the assumption of uniform film growth held well when growth was performed potentiostatically at high potentials (>1.2 V) in the passive domain. As seen in Fig. 41, the spectrum of ε'' for the passive layer showed one major broad peak around 3.6 eV. This $\hat{\varepsilon}$ spectrum for the passive layer was compared by Chen and Cahan with that of α-Fe_2O_3 pellets, as recorded on the same spectroellipsometric system (see Fig. 42). The two ε spectra were found to be quite different. They suggested that the reason for the differences was that the passive film

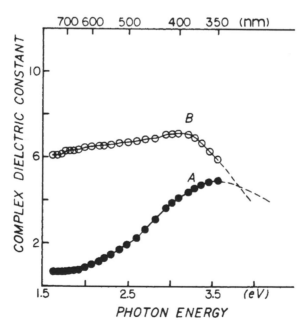

FIG. 41. The spectrum of the complex dielectric constant of the passive film on iron grown in borate buffer at pH 8.4 at 1.35 V, and measured by spectroellipsometry at 1.25 V. The dashed part of the curves is extrapolated using the Lorentz model as a basis. Angle of incidence = 68°. Curve A describes ε'' and curve B-ε'. (From Ref. 18. Reprinted by permission of the publisher, The Electrochemical Society, Inc.)

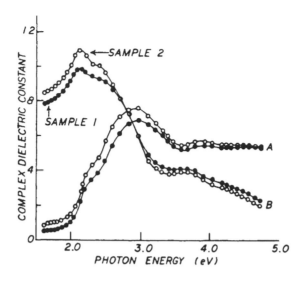

FIG. 42.   The spectrum of the complex dielectric constant of two poly-
crystalline α-Fe_2O_3 powder-pressed pellets sintered at a high temperature, as measured by spectroellipsometry.   Sample 1 sintered at 1100°C
and sample 2 at 1250°C.   A = ε'; B = ε''.   (From Ref. 18.   Reprinted
by permission of the publisher, The Electrochemical Society, Inc.)

is hydrated to some extent, leading to less long-range ordering, compared with crystalline Fe_2O_3, and leading to a shift of the energy of
the 2p-3d transition, due to protonation of O atoms in the hydrous
oxide [18].   According to Chen and Cahan, an optically well-behaved
passive film could be obtained only under potentiostatic growth conditions
at high potentials in the passive domain.   They argued that potentiodynamic experiments, which involve scanning through the domain of
active dissolution, result in substrate roughening simultaneously with
film growth and cause complex optical behavior [18].   The central point
stressed by Chen and Cahan was the optical uniformity of the film as
a function of thickness, which was found experimentally for high potentials in the passive domain.   This leads to the conclusion that the
passive film is a single uniform layer of (hydrated) Fe(III) oxide.
However, based on other experimental evidence, the same authors argued

that gradients in the oxidation state of the Fe ions seem to exist in the passive film at the metal/film and at the film/solution interface. Recently, Jovancicevic et al. [121] reported on a spectroellipsometric investigation of oxide film growth on iron in a similar borate buffer under potential-scanning conditions. Their results show two well-defined passive film development stages, corresponding to two well-resolved voltammetric features recorded between −0.4 and +0.4 V, and ascribed to two gradual steps assigned as a Fe → Fe(II) oxide and Fe(II) oxide → Fe(III) oxide. The reasons for the apparent contradiction between the results of Chen and Cahan [18] and of Jovancicevic et al. [121], seem to be the different potential domains probed and the different experimental conditions employed, i.e., potentiodynamic experiments in a low-potential domain [121] vs. potentiostatic experiments in a high-potential domain [18]. Jovancicevic et al. [121] also reported quite different optical properties for the Fe oxide film, as evaluated from their potential scanning experiments. It is not clear, however, if some effects of substrate roughening did not contribute to the measured ellipsometric effects on which their analysis was based. Still another type of variation of n_f and k_f of the Fe oxide film with applied potential in the same buffer was reported by Kozlowsky and Szklarska-Smialowska [122]. However, the last authors took only measurements of Δ and Ψ at each potential, and therefore the evaluation of their results in terms of n, k, and d for the oxide film at each potential could not be unambiguous. Other recent ellipsometric investigations of passivating oxide layers on metal substrates include the passivating oxides on cobalt [123] and on chromium [124]. Ohtsuka and Sato [123] reported Ψ vs. Δ plots recorded during the cathodic reduction of the passivating oxide film grown on Co in borate buffer. Such plots include two well-defined linear branches. Since similar linear branches appeared also in the experimental Ψ vs. Δ plot during chemical dissolution of the same oxide, the last authors argued that the two linear branches had to be due to two films that dissolve in two well-defined steps—first the external and then the internal layer. Based on the best fit for films, the growth of which would generate these linear segments in the Ψ vs. Δ plot, and assuming thicknesses according to coulombic results and expected stoichiometries, the last authors evaluated for the internal film ("CoO") $\hat{n}_f = 2.3-0.1i$ and for the external film $\hat{n}_f = 3.2-0.56$. These results are unusual in that the internal layer solved for this model exhibits a significantly smaller n_f than does the external layer. Seo et al. [124] measured ellipsometrically the passivating oxide layer on chromium in H_2SO_4 solutions. They found a layer increasing in thickness linearly with applied potential in the passive domain, which reached a maximum thickness of 2.0 nm with optical properties of $\hat{n}_f = 1.80-0.02i$. The authors ascribed these properties to a partly hydrated Cr_2O_3 layer. The work of Kang and Paik [69] on the passive film on nickel in acidic solution is described in detail in Sec. V.A.

Unlike the case of the iron electrode, in which the oxide film can be fully reduced electrochemically, some metal surfaces require mechanical polishing while under cathodic applied potential [61,125,126] or require scraping and transfer to electrolyte under inert atmosphere [127] to maintain an optical reference state of a film-free substrate. Ambrose and Kruger developed the technique of triboellipsometry to allow generation of a film-free surface for metals such as titanium in aqueous electrolytes [125,126]. This technique is based on abrasion of the surface in situ under cathodic applied bias, followed by the retraction of the polishing head, and monitoring of the resulting current and ellipsometric parameters as a function of time under potentiostatic conditions. Ambrose and Kruger suggested that the relative magnitudes of the currents due to metal dissolution and due to the growth of a passivating film could be found from combined ellipsometric and amperometric data under such conditions. The authors reported that the rate of the restoration of passivity, according to triboellipsometric measurements, depended on

the presence of halide ions in solution. It was suggested by Ambrose
and Kruger that the conditions of such experiments simulate the breakdown of passivity by stress-corrosion cracking and passivity restoration
by passive film growth [125,126]. The same method was employed also
by Laser et al., who showed that abrasion in situ under cathodic bias
can generate a film-free metal substrate of good optical quality [61].
The growth of oxide films as thin as 2 nm by linear potential scanning
on a Ti substrate (prepared by abrasion in situ), yielded optical properties very close to those of bulk rutile when analyzed assuming uniform
film growth, as shown in Fig. 43.

Schwager et al. measured ellipsometrically the growth of a surface
film on Li electrodes in propylene carbonate (PC) solutions, following
the preparation of a film-free Li surface by scraping with a scalpel in a
recirculating He atmosphere [127]. Their results are shown in Fig. 44.
Film thicknesses on the Li substrate calculated from ellipsometric measurements were found to increase over periods of several days at open
circuit. These thicknesses were many times larger than those derived
from galvanostatic pulse measurements. This last result is worth attention. The electrical probing of a passivating surface is usually sensitive
to the presence and to the specific thickness of the barrier layer
adjacent to the metal surface. The decay of the anodic dissolution
current or the measured interfacial capacitance will both depend on the
growth and properties of the barrier layer, while both electrical parameters exhibit negligible sensitivity to the presence of thick porous layers
that often form on top of the dense barrier film. On the other hand,
the ellipsometric probe measures both the barrier layer and the thicker
external porous film(s). This is described in Sec. V.A, e.g., for
oxide formation on Fe in alkaline solutions. Film thicknesses derived
from galvanostatic measurements on Li in PC reached 15 nm after 12 days
of immersion, but the ellipsometric measurements showed after the same
time a film of about 150 nm. The last film was best simulated with an

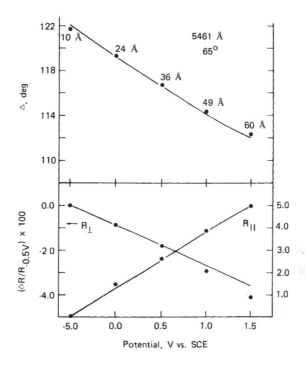

FIG. 43. Experimental and calculated optical changes during thin oxide film growth on a Ti substrate that was polished in situ under cathodic bias. Points, experimental readings; solid lines, the computer solution of a single iostropic film fitted according to all the optical measurements, with $\hat{n}_{film} = 2.43-0.01i$, $\hat{n}_{Ti} = 2.80-3.25i$. (From Ref. 61. Reprinted by permission of the publisher, The Electrochemical Society, Inc.)

assumption of a linear profile of $\hat{n}_f$, changing from 1.57-0.02i down to 1.43-0i in the direction away from the metal-film interface. The authors suggested [127] that reaction with water is the most likely origin of the denser part of the film adjacent to the metal, and according to the value of $\hat{n}_f$ at the beginning of the growth, they concluded that the film could start as a layer of Li_2O rather than LiOH or Li_2CO_3 [127]. They further suggested that the continuing growth of the porous film could indicate a different origin and that precipitation of insoluble products resulting from the decomposition of the solution may be a contributing factor.

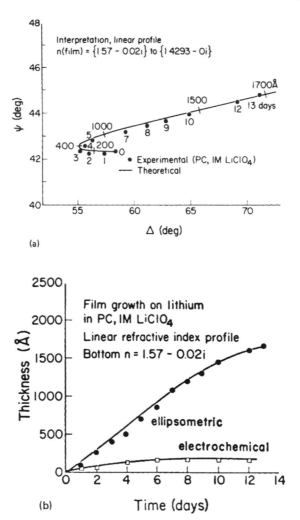

(a)

(b)

FIG. 44. (a) Interpretation of the ellipsometric measurements on Li in 1 M LiClO$_4$ in propylene carbonate with an inhomogeneous film of linear refractive index profile. Thickness of the inhomogeneous film (Å) are given along the computed curve. Period of immersion in days given with measured points. (b) Film growth on Li in 1 M LiClO$_4$ in propylene carbonate derived from ellipsometric measurements for a linear refractive index profile, shown together with film growth as derived from galvanostatic pulse mesurements assuming a static dielectric constant for the film of 4.9. (From Ref. 127. Reprinted by permission of the publisher, The Electrochemical Society, Inc.)

Ritter and Kruger [128,129] have developed a qualitative ellipsometric technique for the study of corrosion processes that take place
under organic coatings. They used as a model iron surfaces coated by
transparent air-cured collodion (cellulose dinitrate) coatings, through
which the coated metal surface sould be examined ellipsometrically while
it was immersed in various electrolytes. The continuous measurements
of Δ and Ψ in a 0.05 M NaCl solution, shown in Fig. 45, revealed oxide
film thinning after 500 hr, followed by uneven metal dissolution, which
brought about optical effects ascribed to metal surface roughening.
Some precipitation causing salt film growth starts taking place subsequently, about 800 hr after the beginning of the experiment. The
assignment of the variations in Δ and Ψ to these sequential processes

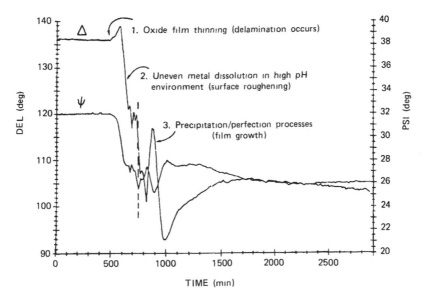

FIG. 45. Δ and Ψ plots vs. time made using an automatic ellipsometer
for an iron-acrylic coating system undergoing corrosion in 0.05 M NaCl
solution. (From Ref. 129. Reprinted with permission from Les Editions
de Physique.)

has been made based particularly on the signs of measured $\delta\Delta$ and $\delta\Psi$. For example, oxide film dissolution brings about an increase in Δ and a decrease in Ψ (opposite to the effects of transparent or semitransparent film growth), and surface roughening causes a lowering in both Ψ and Δ. The microscopic ellipsometric observation with about 10 μm resolution of stainless steel samples undergoing pitting in NaCl solutions [134,135] was described briefly in Sec. III.D. Figure 46 demonstrates the capability of this instrument to provide a topographic map of the surface with 50 μm resolution along the area of the specimen and 0.1 nm in the direction normal to the surface. From such microscopic-ellipsometric investigation, Sugimoto et al. [135] showed (1) that there is a spatial distribution of film thickness within each grain of the 18Cr-18Ni stainless steel specimen, (2) that the first pit occurs in NaCl solutions at a surface site where there is a small variation (0.2—0.5 nm) in film thickness, (3) that the growth of the first pit is followed by film thinning around it and film thickening below it, and (4) that a second pit tends to form on the boundary between the area with the thinned film and the area covered by film of original thickness [135].

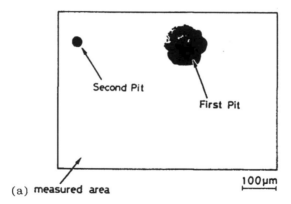

FIG. 46. (a) Optical micrograph of the first and second pits formed at a site on a 18Cr-8Ni stainless steel specimen in 1 M NaCl at 0.8 V. (b) Spatial distribution of passive film thickness in the same surface region, as evaluated with microscopic ellipsometry. (c) Bird's-eye view of the film thickness distribution. (From Ref. 135. Reprinted by permission of the publisher, The Electrochemical Society, Inc.)

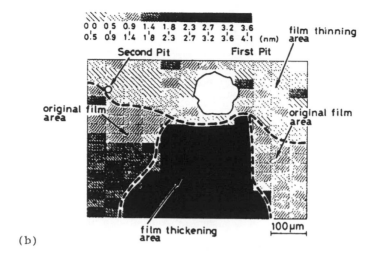

(b)

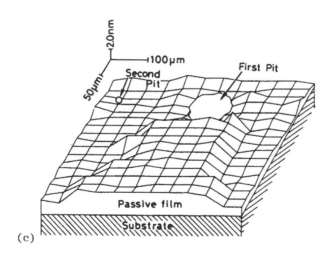

(c)

FIG. 46. Continued

F. Ellipsometric Examination of Semiconductor
Electrode Surfaces

Ellipsometry has been used extensively for the characterization of semiconductor surfaces and of surface film growth on semiconducting substrates [10,104]. Recent contributions demonstrated the power of
spectroellipsometry in providing nondestructive characterization of complex
semiconductor surfaces such as Si surfaces containing crystalline as well
as amorphous Si, SiO_2, and voids [120]. On the other hand, the
application of ellipsometry to the investigation of semiconductor electrode
surfaces has been quite limited. This is despite the fact that the
nature of the surface of the semiconductor electrode and its implications
with respect to protection against corrosion and photocorrosion, as well
as with respect to catalysis in electro- and photoelectrochemical processes, are of great significance. Section V.C contains a detailed
description of the spectroellipsometric work of Heller et al. [105] on
photoelectrochemically deposited Pt films on InP electrodes. Two other
recent contributions, which demonstrate the potential of ellipsometry in
semiconductor electrochemistry, are by Gagnaire et al. [136] and Mercier
et al. [137]. In the first [136], the authors describe an ellipsometric
study of the electrochemical surface modifications of n-InP electrodes.
They demonstrated, with the aid of in situ ellipsometry, that as the InP
electrode is polarized cathodically into the H_2 generation domain, an
indium-rich layer forms on the surface. This layer oxidizes in the
subsequent anodic sweep to form an In oxide layer. The authors used
both a single- and a dual-film model in the interpretation of their
ellipsometric results [136]. Mercier et al. demonstrated by ellipsometry
the uptake of water by SiO_2 films grown on p-Si [137]. They showed
film thickness increase due to immersion in an aqueous solution when the
applied voltage was much smaller than the voltage of oxide film growth
(200 V) or even when no voltage was applied at all. The authors concluded from the ellipsometric investigation that further growth of SiO_2
films takes place in the presence of water molecules and also showed

that these thickness variations cannot be properly determined by measurements of the capacitance because the uptake of water affects the dielectric properties of the layer. This is another example where film thickness probing by ellipsometry proves to be more reliable than probing by electrical or electrochemical methods. Finally, some applications of ellipsometry to semiconductor technology areas where electrochemical (or chemical) processing is employed will be briefly reviewed. Aspnes et al. [138] reported spectroellipsometric studies of anodically grown oxide films on GaAs under galvanostatic conditions in phosphoric acid/glycol-based electrolytes. They stressed the advantage of optical nondestructive studies for the characterization of such layers, compared with photoemission or Auger profiling, which are suspect of electron and ion beam reduction artifacts. Their spectroscopic measurements could show that a layer modeled as a-As at the GaAs-oxide interface is 3—10 Å thick, depending on the galvanostatic growth current. They could also show that an absorption component in the nominally transparent energy range for the oxide (< 4.5 eV) was present, equivalent to 0.24—0.99% by volume of a-As, depending on the electrochemical growth conditions. Finally, two recent contributions by Palik et al. [139,140] describe ellipsometric measurements during chemical and electrochemical etching of Si in KOH solutions. A thin "oxide-like layer" has been identified ellipsometrically by the authors as the reason for the etch-stop phenomenon found for p-Si in KOH [139]. Measured changes in the thickness and stoichiometry of the surface SiO_x phase during various treatments provided a basis for a model of the Si etching chemistry, and for the orientation dependence of the etch rate [140].

VI.   OUTLOOK

The recent interest in the application of ellipsometry for characterizing electrochemical interfaces is a result of two developments. First, advances in instrumentation have led to replacement of the slow manual

ellipsometer by the automated ellipsometer capable of λ scanning and of
time-resolved measurements of interfacial dynamics. Second, interest in
the characterization of surface films, in terms of thickness, microstructure, and electronic properties, has recently widened. This interest is
now shared not only by electrochemists working on metal passivation or
on oxide films, but also by the group of electrochemists working on
various new types of filmed electrodes ("modified electrodes"). The
need for better understanding of the physical properties and of the
electrochemical dynamics of conducting polymer films, or of polymer
films with incorporated redox components, is likely to add new users
of ellipsometry within the electrochemical community. The ability to
obtain microstructural information on composite films in situ is a recently
developed attractive feature of spectroellipsometry that is likely to be
used more extensively in the future. Characterization by ellipsometry
of surface films on semiconductor electrodes is also a field that is likely
to expand. Recent results of spectroellipsometric investigations of
complex semiconductor surfaces in dry systems will encourage others to
attempt similar studies of semiconductor surfaces in electrochemical
environments. As for future extensions of instrumental capabilities,
recent demonstrations of submicrosecond time resolution [133], 10-μm
spatial resolution [135], and possible applications in the IR region [143],
promise to catalyze a significant follow-up. Finally, like any other
method for the characterization of the electrochemical interface, ellipsometry provides information that can and should be complemented by
other techniques. A suggestion for future work, mentioned above
Sec. V), is to combine in situ ellipsometric and quartz-crystal-micro-
balance measurements. Another example is to complement the information
on the electronic, dimensional, and microstructural properties of a
surface film, as derived from in situ spectroellipsometry in the UV-
visible, with spectroscopic information on the molecular nature of components of the film. The recent developments in ellipsometry have

provided a good basis for its becoming an important component in future "packages" of instrumental techniques that will be applied for the in situ characterization of electrode-solution interfaces.

ACKNOWLEDGMENT

I wish to thank all of my colleagues in the Department of Chemistry, University of Tel Aviv, Israel, and in the R&D group of the Mechanical and Electronic Engineering Division of Los Alamos National Laboratory, United States, who have collaborated with me in ellipsometric investigations of electrochemical systems. Special thanks are due to A. Redondo of Los Alamos for his role in the ellipsometric work performed here, for stimulating discussions, and for generating the simulated plots found in Sec. IV.B. Thanks are due to the Department of Energy, Office of Energy Storage and Distribution, for partial support during the time this chapter was written.

REFERENCES

1. R. M. A. Azzam and N. M. Bashara, *Ellipsometry and Polarized Light*, North-Holland, Amsterdam, 1977.

2. Proceedings of the Fourth International Conference on Ellipsometry, Surf. Sci. *96* (1980).

3. Proceedings of the Fifth International Conference on Ellipsometry, J. de Physique Paris Colloq. *C10* (1983).

4. R. H. Muller, in *Advances in Electrochemistry and Electrochemical Engineering* (R. H. Muller, ed.), Wiley, New York, 1973, pp. 168–226.

5. J. Kruger, in *Advances in Electrochemistry and Electrochemical Engineering* (R. H. Muller, Ed.), Wiley, New York, 1973, pp. 227–280.

6. W. Paik, in *MTP International Review of Science*, Physical Chemistry Series One (J. O'M. Bockris, ed.), Butterworth, London, 1973, Vol. 6, p. 239.

7. R. Greef, in *Comprehensive Treatise of Electrochemistry* (J. O'M. Bockris, B. E. Conway, and E. Yeager, eds.), Plenum, New York, 1984, Vol. 8, p. 339.

8. F. L. McCrackin and J. Colson, *National Bureau of Standards Technical Note,* No. 242 (1964).

9. W. Paik and J. O'M. Bockris, Surf. Sci. *28:*61 (1971).

10. F. Hottier and J. B. Theeten, J. Cryst. Growth *48:*644 (1980).

11. S. Gottesfeld and S. Srinivasan, J. Electroanal. Chem. *86:*89 (1978).

12. J. L. Ord, J. Electrochem. Soc. *129:*335 (1982).

13. J. A. Johnson and N. M. Bashara, J. Opt. Soc. Am. *61:*457 (1971).

14. B. D. Cahan and R. F. Spanier, Surf. Sci. *16:*166 (1969).

15. D. E. Aspnes, Opt. Commun. *8:*122 (1973).

16. P. S. Hauge and F. H. Dill, IBM J. Res. Dev. *17:*472 (1973).

17. J. Rishpon, I. Reshef, and S. Gottesfeld, Anal. Instrum. 14(2): 105 (1985).

18. C. T. Chen and B. D. Cahan, J. Electrochem. Soc. *129:*17 (1982).

19. J. Rishpon and S. Gottesfeld, J. Electrochem. Soc. *131:*1960 (1984).

20. D. E. Aspnes and A. A. Studna, Appl. Opt. *14:*220 (1975).

21. D. E. Aspnes, J. Opt. Soc. Am. *64:*639 (1974).

22. R. H. Muller and H. J. Mathieu, Appl. Opt. *13:*2222 (1974).

23. R. H. Muller and J. C. Farmer, Rev. Sci. Instrum. *55:*371 (1984).

24. S. V. Pihlajamaki and J. J. Kankare, J. Electroanal. Chem. (1984).

25. P. J. Hyde, C. J. Maggiore, A. Redondo, S. Srinivasan, and S. Gottesfeld, J. Electroanal. Chem. *186:*267 (1985).

26. S. N. Jasperson, D. K. Burge, and R. C. O'Handley, Surf. Sci. *37:*548 (1973).

27. R. C. O'Handley, D. K. Burge, S. N. Jasperson, and E. J. Ashley, Surf. Sci. *50:*407 (1975).

28. V. M. Bermudez and V. H. Ritz, Appl. Opt. *17:*542 (1978).

29. G. A. Candela and D. Chandler-Horowitz, *SPIE,* Vol. *480,* Integrated Circuit Metrology II, p. 1 (1984).

30. A. Redondo and S. Gottesfeld, unpublished results.

31. D. Chandler-Horowitz and G. A. Candela, J. Phys. Paris Colloq. *C10:*23 (1983).

32. R. C. O'Handley, J. Opt. Soc. Am. *63:*523 (1973).

33. A. Moritani, Y. Okuda, and J. Nakai, Appl. Opt. *22:*1329 (1983).

34. A. Moritani and C. Hamaguchi, Appl. Phys. Lett. *46:*746 (1985).

35. B. D. Cahan, J. Horkans, and E. Yeager, Surf. Sci. *37:*559 (1973).

36. S. Gottesfeld and B. Reichman, Surf. Sci. 44:377 (1974).

37. S. Gottesfeld, M. Babai, and B. Reichman, Surf. Sci. 56:373 (1976).

38. S. Gottesfeld, M. Babai, and B. Reichman, Surf. Sci. 57:251 (1976).

39. S. Gottesfeld and B. Reichman, J. Electroanal. Chem. 67:169 (1976).

40. D. P. Arndt, Appl. Opt. 23:3571 (1984).

41. D. E. Aspnes and H. G. Craighead, Appl. Op. 25:1299 (1986).

42. K. D. Nagele and W. J. Plieth, Electrochim. Acta 25:241 (1980).

43. F. Wooten, Optical Properties of Solids, Academic, New York, 1972. For a detailed updated source of optical constants of solids, see Handbook of Optical Constants of Solids (E. D. Palik, ed.), Academic, Orlando, Fla., 1985.

44. S. Gottesfeld, M. T. Paffett, and A. Redondo, J. Electroanal. Chem. 205:163 (1986).

45. J. M. Otten and W. Visscher, J. Electroanal. Chem. 55:1, 13 (1974).

46. D. J. DeSmet and J. L. Ord, J. Electrochem. Soc. 130:280 (1983).

47. J. L. Ord and W. P. Wang, J. Electrochem. Soc. 130:1809 (1983).

48. J. C. Clayton and D. J. DeSmet, J. Electrochem. Soc. 123:174 (1976).

49. M. A. Hopper and J. L. Ord, J. Electrochem. Soc. 120:183 (1973).

50. M. Babai and S. Gottesfeld, Surf. Sci. 96:461 (1980).

51. J. L. Ord, J. Electrochem. Soc. 129:335 (1982).

52. D. E. Aspnes, E. Kinsbron, and D. D. Bacon, Phys. Rev. B21: 3290 (1980).

53. Z. Q. Huang and J. L. Ord, J. Electrochem. Soc. 132:24 (1985).

54. D. Michell, D. A. J. Rand, and R. Woods, J. Electroanal. Chem. 84:117 (1977).

55. C. M. Carlin, L. J. Kepley, and A. J. Bard, J. Electrochem. Soc. 132:353 (1985).

56. S. Gottesfeld, A. Redondo, and S. W. Feldberg, Abstract 507 in Extended Abstracts, Electrochem. Soc. Meet., 86(2) (1986).

57. O. Melroy, K. Kanazawa, J. G. Gordon II, and D. Buttry, Langmuir 2:697 (1986).

58. A. Hamnet and A. R. Hillman, Ber. Bunsenges. Phys. Chem. 91(4): 329 (1987).

59. C. G. Matthews, J. L. Ord, and W. P. Wang, J. Electrochem. Soc. 130:285 (1983).

60. J. L. Ord, J. Electrochem. Soc. *129:*767 (1982).

61. D. Laser, M. Yaniv, and S. Gottesfeld, J. Electrochem. Soc. *125:* 358 (1978).

62. S. Shibata and M. P. Sumino, Electrochim. Acta *17:*2215 (1972).

63. S. Gottesfeld, M. Yaniv, D. Laser, and S. Srinivasan, J. Phys. Paris Colloq. *C5:*145 (1977).

64. J. Horkans, B. D. Cahan, and E. Yeager, Surf. Sci. *46:*1 (1974).

65. M. Peuckert and H. P. Bonzel, Surf. Sci. *145:*239 (1984).

66. O. A. Albani, J. O. Zerbino, J. R. Vilche, and A. J. Arvia, Electrochim. Acta *31:*1403 (1986).

67. T. Ohtsuka, K. Schroner, and K. E. Heusler, J. Electroanal. Chem. *93:*171 (1978).

68. W. Visscher and E. Barendrecht, Surf. Sci. *135:*436 (1983).

69. Y. Kang and W. Paik, Surf. Sci. *182:*257 (1987).

70. See, e.g., B. MacDougall and M. Cohen, J. Electrochem. Soc. *121:*1152 (1974).

71. J. O'M. Bockris, A. K. N. Reddy, and B. Rao, J. Electrochem. Soc. *113:*1133 (1966).

72. A. K. N. Reddy, and B. Rao, Can. J. Chem. *47:*2688 (1969).

73. J. J. Caroll, T. E. Madey, A. J. Melmed, and D. R. Sandstrom, Surf. Sci. *96:*508 (1980).

74. L. Bohn, in *Polymer Handbook* (2nd Ed.) (J. Brandrup and E. H. Immergut, eds.), Wiley, New York, 1975, pp. III−241.

75. H. Arwin, D. E. Aspnes, R. Bjorklund, and I. Lundström, Synth. Met. *6:*309 (1983).

76. S. Gottesfeld, A. Redondo, and S. W. Feldberg, J. Electrochem. Soc. *134:*271 (1987).

77. S. Gottesfeld and J. D. E. McIntyre, J. Electrochem. Soc. *126:*742 (1979).

78. S. Gottesfeld, J. Electrochem. Soc. *127:*272 (1980).

79. T. Kobayashi, H. Yoneyama, and H. Tamara, J. Electroanal. Chem. *177:*281 (1984).

80. D. J. DeSmet, Electrochim. Acta *21:*1137 (1976).

81. F. H. P. M. Habraken, O. L. J. Gijzeman, and G. A. Bootsma, Surf. Sci. *96:*482 (1980).

82. E. C. Butcher, A. J. Dyer, and N. E. Gilbert, Br. J. Appl. Phys. *1:*1673 (1968).

83. P. J. Powell and W. E. Spicer, Phys. Rev. *2*: 2183 (1970).

84. W. J. Plieth, J. Phys. Paris Colloq. *C5*: 215 (1977).

85. W. J. Plieth, Isr. J. Chem. *18*: 105 (1978).

86. C. Strachan, Proc. Cambridge Philos. Soc. *29*: 116 (1933).

87. G. A. Bootsma and F. Meyer, Surf. Sci. *14*: 52 (1969).

88. D. V. Sivukhin, Sov. Phys. JETP *3*: 269 (1956).

89. T, Smith, J. Opt. Soc. Am. *58*: 1069 (1968).

90. R. M. Azzam and N. M. Bashara, J. Opt. Soc. Am. *64*: 128 (1974).

91. M. Elshazly-Zaghloul, R. M. A. Azzam, and N. M. Bashara, Surf. Sci. *56*: 281 (1976).

92. D. den Engelsen, and B. de Koning, J. Chem. Soc. Faraday Trans. 1 *70*: 1603 (1974).

93. G. T. Ayoub and N. M. Bashara, J. Opt. Soc. Am. *68*: 978 (1978).

94. P. J. Hyde and S. Gottesfeld, Surf. Sci. *149*: 601 (1985).

95. M. Fleischman and I. R. Hill, in *Surface Enahnced Raman Scattering* (R. K. Chang and T. F. Furtak, eds.), Plenum, New York, 1982, p. 275.

96. J. K. Foley, C. Koreniewski, J. L. Daschbach, and S. Pons, in *Electroanalytical Chemistry* (A. J. Bard, ed.), Dekker, New York, 1986, Vol. 14.

97. A. Bewick and J. Robinson, Surf. Sci. *55*: 349 (1976).

98. A. Bewick and J. Robinson, J. Electroanal. Chem. *60*: 163 (1975).

99. A. Tadjeddine, J. Electroanal. Chem. *169*: 129 (1984).

100. Y. C. Chin and M. A. Genshaw, J. Phys. Chem. *73*: 3571 (1969).

101. W. Paik, M. A. Genshaw, and J. O'M. Bockris, J. Phys. Chem. *74*: 4266 (1970).

102. E. Bertel, K. Schwaha, and N. P. Netzer, Surf. Sci. *83*: 439 (1979).

103. O. Hunderi, Surf. Sci. *96*: 1 (1980).

104. D. E. Aspnes, Thin Solid Films, *89*: 249 (1982).

105. A. Heller, D. E. Aspnes, J. D. Porter, T. T. Sheng, and R. G. Vadimsky, J. Phys. Chem. *89*: 4444 (1985).

106. S. Gottesfeld, M. T. Paffett, and A. Redondo, J. Electroanal. Chem. *205*: 163 (1986).

107. K. S. Kang and J. L. Shay, J. Electrochem. Soc. *130*: 766 (1983).

108. R. H. Muller and J. C. Farmer, Surf. Sci. *135:*521 (1983).

109. F. Chao and M. Costa, Surf. Sci. *135:*497 (1983).

110. D. M. Kolb and R. Kötz, Surf. Sci. *64:*698 (1977).

111. D. M. Kolb, in *Advances in Electrochemistry and Electrochemical Engineering* (H. Gerischer and C. W. Tobigo, eds.), Wiley, New York, 1978, Vol. 11, p. 125.

112. J. D. E. McIntyre and D. M. Kolb, Symp. Faraday Soc. *4:*99 (1976).

113. D. M. Kolb, D. Leutloff, and M. Przasnyski, Surf. Sci. *47:*622 (1975).

114. M. J. Dignam and M. Moskovitz, J. Chem. Soc. Faraday Trans. 2 *69:*56 (1973).

115. M. J. Dignam, M. Moskovitz, and R. W. Stobie, Trans. Faraday Soc. *67:*3306 (1971).

116. T. Takamura, F. Watanabe, and K. Takamura, Electrochim. Acta *19:933 (1974).*

117. M. Y. Abyaneh, W. Visscher, and E. Barendrecht, Electrochim. Acta *28:*285 (1983).

118. J. C. Farmer and R. H. Muller, J. Electrochem. Soc. *132:*313 (1985).

119. R. H. Muller and C. G. Smith, Surf. Sci. *96:*375 (1980).

120. R. W. Collins, B. G. Yacobi, K. M. Jones, and Y. S. Tsuo, J. Vac. Sci. Technol. *A4:*153 (1986).

121. V. Jovancicevic, R. Kainthla, Z. Tang, B. Yang, and J. O'M. Bockris, Langmuir *3:*338 (1987).

122. Z. Szklarska-Smialowska and W. Kozlowski, J. Electrochem. Soc. *131:*234 (1984).

123. T. Ohtsuka and N. Sato, J. Electrochem. Soc. *128:*2522 (1981).

124. M. Seo, R. Saito, and N. Sato, J. Electrochem. Soc. *127:*1909 (1980).

125. J. R. Ambrose and J. Kruger, Corrosion *28:*30 (1972).

126. J. R. Ambrose and J. Kruger, J. Electrochem. Soc. *121:*599 (1974).

127. F. Schwager, Y. Geronov, and R. H. Muller, J. Electrochem. Soc. *132:*285 (1985).

128. J. J. Ritter and J. Kruger, Surf. Sci. *96:*364 (1980).

129. J. J. Ritter and J. Kruger, J. Phys. Paris Colloq. *C10:*225 (1983).

130. Y. Bouziem, F. Chao, M. Costa, A. Tadjeddine, and G. Lecayon, J. Electroanal. Chem. *172:*101 (1984).

131. F. Chao, M. Costa, P. Lang, and E. Lheritier, J. Electroanal. Chem. *162:*814 (1985).

132. J. L. Bredas, B. Themans, J. G. Fripiat, J. M. Andre, and R. R. Chance, Phys. Rev. *B29:*6771 (1984).

133. G. E. Jellison and D. H. Lowndes, Appl. Opt. *24:*2948 (1985).

134. K. Sugimoto and S. Matsuda, J. Electrochem. Soc. *130:*2323 (1983).

135. K. Sugimoto, S. Matsuda, Y. Ogiwara, and K. Kitamura, J. Electrochem. Soc. *132:*1791 (1985).

136. A. Gagnaire, J. Joseph, A. Etchberry, and J. Gautron, J. Electrochem. Soc. *132:*1655 (1985).

137. J. J. Mercier, F. Fransen, F. Cardon, M. J. Madou, and W. P. Gomes, Ber. Bunsenges. Phys. Chem. *89:*117 (1985).

138. D. E. Aspnes, G. P. Schwartz, G. J. Gualtieri, A. A. Studna, and B. Schwartz, J. Electrochem. Soc. *128:*290 (1981).

139. E. D. Palik, V. M. Bermudez, and O. J. Glembocki, J. Electrochem. Soc. *132:*135 (1985).

140. E. D. Palik, V. M. Bermudez, and O. J. Glembocki, J. Electrochem. Soc. *132:*871 (1985).

141. Z. Szklarska-Smialowska, T. Zakroczymski, and C. J. Fan, J. Electrochem Soc. *132:*2543 (1985).

142. G. N. Salaita, D. A. Stern, F. Lu, H. Baltruschat, B. C. Shardt, J. L. Stickney, M. P. Soriaga, D. G. Frank, and A. T. Hubbard, Langmuir *2:*828 (1986).

143. R. T. Graf, J. L. Koenig, and H. Ishida, Anal. Chem. *58:*64 (1986).

144. R. Steiger, Helv. Chim. Acta *54:*2645 (1971).

145. W. Knoll, J. Rabe, M. R. Philpott, and J. D. Swolen, Thin Solid Films *99:*173 (1983).

146. D. L. Allara and R. G. Nuzzo, Langmuir *1:*45 (1985).

VOLTAMMETRY AT ULTRAMICROELECTRODES

R. Mark Wightman and David O. Wipf

Department of Chemistry
Indiana University
Bloomington, Indiana

I. INTRODUCTION

This review deals with the properties of voltammetric electrodes of very
small dimensions. When the dimensions of an electrode are decreased
from the millimeter to the micrometer scale, many changes occur in the
voltammetric behavior. Although the advantageous properties of very
small electrodes were recognized for many years, research in this area
did not become very active until the late 1970s. The advances made in
the fields of electronics, especially in the mesurement of very small
currents, and the advent of microstructural materials, provided the
tools that were necessary to build and use electrodes of small dimensions.
Fleischmann and co-workers at the University of Southhampton initiated
much of this activity with an interest in understanding electrode mechanisms under conditions of high current density [1]. At the same time,
miniature electrodes were being developed for use as in vivo probes
[2,3]. By 1981 it was clear that many new areas which were inaccess-
ible to electrodes of larger size could be explored with very small
electrodes, and an early review summarizes many of these expectations
[4]. Since that review appeared, significant advances have been made
in the exploitation of the many aspects of these small electrodes.

This area is still sufficiently new that a uniform nomenclature for
these electrodes has not yet been developed. Initially, we referred to
these electrodes as microvoltammetric electrodes [4]. The term "ultramicroelectrode" seemed inappropriate since many of the phenomena
discussed in this review can be observed at electrodes that are large in
comparison to the electrodes that electrophysiologists routinely use.
The term "microelectrode" is already in routine use for electrodes with
dimensions approaching a centimeter or greater. However, the term
"ultramicroelectrodes" is seeing increased use today and is the term
used to describe these small electrodes in this chapter.

Ultramicroelectrodes have seen increased use because they offer dramatic improvements in the quality of electrochemical data. Depending on the experimental conditions, these include increased temporal resolution, increased current density, and decreased sensitivity to the effects of solution resistance. In addition, previously impossible experiments have been made possible with ultramicroelectrodes. New domains that have been explored include submicrosecond electrochemistry [5,6], electrochemistry in gases [7], and electrochemistry under time-independent conditions [8]. However, to be able to explore these new domains, the theory and methods of construction of these small electrodes need to be considered.

The electrochemical responses at ultramicroelectrodes can differ greatly from those seen at conventional microelectrodes. This is because, when operated on a time scale of seconds, the dimensions of the electrode are much smaller than the diffusion distance for molecules in solution. This deviation from planar diffusion leads to the steady-state response often observed at ultramicroelectrodes. This type of behavior was anticipated for many years, but its predominance in experiments with ultramicroelectrodes has led to a resurgence of thinking about diffusion to these microstructures. Because this is such a predominant feature of the response at ultramicroelectrodes, this area will be reviewed in some detail. As will be shown, complete descriptions of the diffusional based response for a wide number of geometries of ultramicroelectrodes now exist.

The unique features of ultramicroelectrodes assure that they will continue to be important in electrochemical studies. The small size, reduced ohmic drop, increased temporal resolution, and steady-state diffusional profiles all can be used to advantage in a number of applications. As will be seen, significant progress has been made in the realization of the many new domains made available by ultramicroelectrodes.

II. DIFFUSION AT ULTRAMICROELECTRODES

When electrolysis occurs, a concentration gradient is generated between
the electrode surface and solution [9]. For this reason the current
measured at all voltammetric electrodes depends on the modes of mass
transport that are operant. At ultramicroelectrodes the predominant
mode is diffusion. Because the flux as a result of diffusion is very
large, the effects of convection tend to be less apparent than at electrodes of conventional size. For this reason, the majority of the theoretical treatments of currents at ultramicroelectrodes have dealt with
diffusion-limited processes. The magnitude and time dependence of the
diffusion processes are determined in part by the geometry of the
electrode. However, because we are considering structures of very
small dimensions, the conditions of semi-infinite diffusion are assumed
in the majority of these treatments.

In the following discussion, each electrode geometry will first be
considered for the case of chronoamperometry. In the simplest form of
this experiment, the potential is stepped from a region where electrolysis
does not occur to a value where the current is limited only by diffusion
of species to the electrode. Under these conditions, derivation of the
current-time response requires solution of Fick's laws of diffusion under
the appropriate boundary conditions. The other case considered is
cyclic voltammetry. Analytical solutions for this case tend to be more
complicated because the voltage dependence of the current must be
evaluated as well as diffusion processes. However, as shown, the
chronoamperometric response in many cases tends to be similar to the
limiting current obtained in cyclic voltammetry. Although this section
of the review will consider only diffusion-controlled, reversible systems,
these expressions are important because they serve as a starting place
to evaluate various other factors that can affect the electrochemical
response. These include the effects of finite kinetics of interfacial
electron transfer and chemical reactions that accompany the electron-
transfer reaction.

A. Diffusion at Spherical Electrodes

The processes that occur at spherical electrodes during electrolysis exemplify many of the factors which are seen at other geometries of ultramicroelectrodes. The diffusion problems were solved many years ago because of the importance of spherical electrodes in the classical polarographic experiment with a dropping mercury electrode [10]. As described later in this chapter, hemispherical microelectrodes of micrometer dimensions are simple to fabricate and use. Thus these equations are of experimental utility today. For both of these reasons, it is worth considering diffusion properties during electrolysis with spherical electrodes in some detail.

We consider first a chronoamperometric experiment at an electrode with the geometry of a sphere of radius r_0. The problem will be considered in spherical coordinates, where r is the radial distance from the center of the electrode. Only one vector from the electrode needs to be considered because the diffusion field is symmetric, and the result will be the same over all angular coordinates. The boundary conditions for the reduction of species O, of bulk concentration C_0^*, are

$$\lim_{r \to \infty} C_0(r,t) = C_0^*$$

$$C_0(r,0) = C_0^* \qquad (r > r_0)$$

$$C_0(r_0,t) = 0 \qquad (t > 0)$$

To evaluate the faradaic current (i), we need a solution to the gradient of concentration at the electrode surface:

$$\frac{i}{nFA} = D_0 \left[\frac{\partial C_0(r_0,t)}{\partial r} \right] \tag{1}$$

where D_0 is the diffusion coefficient of the species being electrolyzed, A the area of the electrode, F the Faraday, and n the number of electrons per molecule oxidized or reduced. To obtain a solution for the concentration gradient, Fick's second law must be evaluated. For a spherical coordinate system, this is written as

$$\frac{\partial C_o(r,t)}{\partial t} = D_o \left[\frac{\partial^2 C_o(r,t)}{\partial r^2} + \frac{(2/r)\partial C_o(r,t)}{\partial r} \right] \tag{2}$$

Equation (2) can be simplified if we set $v(r,t) = rC_o(r,t)$. Substitution then gives

$$\frac{\partial v(r,t)}{\partial t} = D_o \left[\frac{\partial^2 v(r,t)}{\partial r^2} \right] \tag{3}$$

Solution of this second-order differential equation, with the appropriate boundary conditions, gives the expression for the concentration profile [11]:

$$C_o(r,t) = C_o^* \left(1 - \frac{r_0}{r} \right) \text{erfc} \left[\frac{r - r_0}{(4D_o t)^{1/2}} \right] \tag{4}$$

When the dimensions of the electrode are large, i.e., the radius of curvature, r_0, is infinite, then

$$C_o(r,t) = C_o^* \text{ erf} \left[\frac{x}{(4D_o t)^{1/2}} \right] \tag{5}$$

where $x = r - r_0$. This is the expression for the concentration profile at large, planar electrodes. On the other hand, at long times, then [11]

$$C_o(r, t \rightarrow \infty) = C_o^* \left(1 - \frac{r_0}{r} \right) \tag{6}$$

A graphical comparison of the concentration profiles at large and small spherical electrodes is given in Fig. 1. The equation for planar diffusion predicts that the concentration profile will keep extending into solution as a function of time. At a small spherical electrode the growth of the region perturbed by electrolysis, the diffusion layer, is initially similar to that observed at a planar electrode, but at longer times the growth slows considerably.

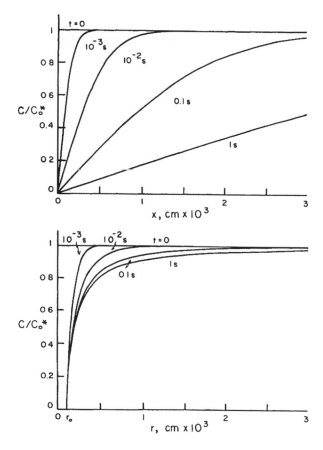

FIG. 1.   Concentration profiles at large ($r_0 \to \infty$, top) and small ($r_0 = 10^{-4}$ cm, bottom) for different times after the application of a potential step of sufficient magnitude to drive the concentration of the electro-active species to zero at the electrode surface.   Evaluated for $D_0 = 10^{-5}$ cm^2 sec^{-1}.

The reason for this behavior can be seen in another way if one considers a different representation of the diffusion layer.   At short times the dimensions of the diffusion layer are smaller than that of the electrode, and a molecule at the farthest edge of the diffusion layer "sees" the electrode as a giant plane (Fig. 2).   However, when the dimensions of the diffusion layer exceed that of the electrode, the

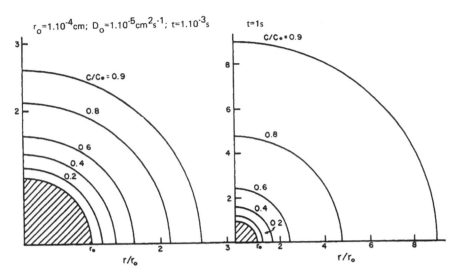

FIG. 2.   Concentration profiles for a spherical electrode at 10^{-3} sec
(left) and 1 sec (right) after the application of a potential step of
sufficient magnitude to drive the concentration of the electroactive
species to zero at the electrode surface.

spherical nature of the experiment becomes more apparent.   The population of molecules "feeding" the electrode comes from a solid angle that
contains an ever-increasing volume.   Thus the number of electroactive
molecules which have access to a spherical electrode exceeds that of a
planar electrode.

Differentiation of Eq. (4) with respect to r and integration of the
result over the area of the electrode gives the current at a spherical
electrode

$$i = nFAD_oC_o^* \left[\frac{1}{(\pi D_o t)^{1/2}} + \frac{1}{r_0} \right] \tag{7}$$

where A is given by $4\pi r_0^2$ for a sphere and $2\pi r_0^2$ for a hemisphere.
At short times the currents at a spherical electrode are identical to those
found under conditions of planar diffusion.   However, at longer times
the time-dependent term becomes negligible and the expression converges

to a constant value as illustrated in Fig. 3. As we shall see, Eq. (7) is typical of all geometries of ultramicroelectrodes—the response differs from the results obtained at electrodes large enough to be considered planar by additional terms that reflect the larger number of molecules that have access to the electrode. Of particular interest is the time-independent nature of the second term for spherical geometries.

For practical use it is convenient to have guideline figures to predict when the current will be predominantly steady state in nature. This can be achieved by division of Eq. (7) by the current at a planar electrode, the Cottrell equation [9]. If the areas are considered equal, then

$$\frac{i_{sphere}}{i_{plane}} = 1 + \frac{(\pi D_o t)^{1/2}}{r_0} \tag{8}$$

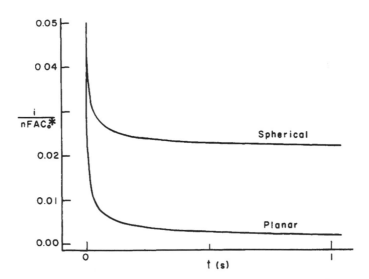

FIG. 3. Current vs. time trace for a potential-step experiment at planar and spherical electrodes, evaluated for $D_0 = 10^{-5}$ cm^2 sec^{-1} and $r_0 = 5 \times 10^{-4}$ cm.

It can be seen from this relationship that the dimensionles parameter, $D_0 t/r_0^2$, provides the desired guideline. If $D_0 t/r_0^2$ has a value of 32, the time-independent part of Eq. (7) will be 10 times larger than the Cottrellian portion. For aqueous solutions, where $D_0 \sim 1 \times 10^{-5}$ cm^2/sec for small molecules, this corresponds to a time of 0.25 sec for a typical ultramicroelectrode radius of 5×10^{-4} cm. For the current to have only a 10% contribution from the time-independent term, $D_0 t/r_0^2$ must have a value of 3×10^{-3}. This corresponds to a time of 0.08 msec. Thus over most accessible time scales, some effect of the time-independent term can be seen with a spherical ultramicroelectrode of these dimensions [12].

The response predicted by Eq. (7) was experimentally verified by Shain and Martin with the sid of a hanging mercury drop electrode [13]. An immediate use of Eq. (7) is for the determination of diffusion coefficients. A plot of the measured current (i) against $t^{-1/2}$ should be linear. For a sphere this straight line will have the following slope and intercept:

$$\text{slope} = 4\pi^{1/2} r_0^2 nFC_o^* D_o^{1/2} \tag{9}$$

$$\text{intercept} = 4\pi r_0 nFC_o^* D_o \tag{10}$$

Thus if r_0 and C_o^* are known, n and D_0 can be determined in one experiment. This approach has been used to determine the diffusion coefficients of several metal ions in aqueous solutions [14]; however, the relatively large size of the sphere employed required long electrolysis times. This type of experiment is more practical today with the advent of ultramicroelectrodes.

In any controlled-voltage experiment, conducted with a large value of $D_0 t/r_0^2$, the limiting current under diffusion-controlled conditions will be the same, whatever the potential excitation. For example, with cyclic voltammetry at sufficiently slow scan rates, the limiting current (i_ℓ) is that predicted by Eq. (10). This is because the limiting current is

reached at very long times for a sufficiently slow scan rate. This is
shown for the limiting current at a hemispherical mercury electrode
($r_0 = 5.5 \times 10^{-4}$ cm) for the reduction of $Ru(NH_3)_6^{3+}$ in aqueous solution (Fig. 4) which is given by

$$i_\ell = 2 \pi r_0 nFD_o C_o^*$$ (11)

just as would be predicted for the long-time result in chronoamperometry
[15]. The reverse wave is not observed under these steady-state conditions because the electrolysis product leaves the diffusion layer at an
enhanced rate for the same reasons the inward flux is so large.

    If ultramicroelectrodes are used at short time scales (small values of
$D_o t/r_0^2$), the voltammograms resemble those observed at planar electrodes. This is also illustrated for the same electrode in the same
solution in Fig. 4. The voltammogram is peak shaped at high scan rates,
and a wave exists on the reverse scan for the oxidation of products
generated on the forward scan.

    These expectations for the shape of cyclic voltammograms, based on
an understanding of diffusion-controlled processes at spherical electrodes,
have a rigorous basis. In their classic work on cyclic voltammetry,
Nicholson and Shain published numerical tables with values for what
they termed the "spherical correction" [16]. To predict the current
at a spherical, electrode, one simply weights the values in the table in
the appropriate way and the complete voltammogram can be calculated.
These values are invaluable for predictive purposes when the use of
spherical ultramicroelectrodes is considered.

    The steady-state nature of cyclic voltammograms at ultramicroelec-
trodes has been used to advantage in several applications, as we will
see later in this chapter. However, one feature that is of particular
interest is the ease with which the shape of a voltammogram can be
evaluated when the curve is steady state. Under steady-state conditions
[Eq. (1)] the derivative of the concentration with respect to time must

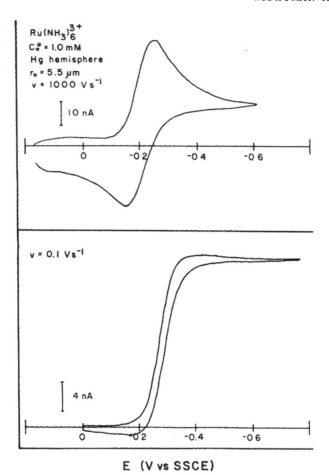

FIG. 4.   Cyclic voltammograms for reduction of 1 mM $Ru(NH_3)_6^{3+}$ in an aqueous 0.1 M sodium trifluoroacetate solution at a hemispherical mercury ultramicroelectrode (r_0 = 5.5 μm) at slow and fast scan rates.   SSCE = saturated sodium calomel electrode.

TABLE 1

Current Functions $\sqrt{\pi}\chi(\sigma t)$ for Reversible Charge Transfer[a]

$(E - E_{1/2})n$[b] (mV)	$\sqrt{\pi}\chi(\sigma t)$	$\phi(\sigma t)$	$(E - E_{1/2})n$[b] (mV)	$\sqrt{\pi}\chi(\sigma t)$	$\phi(\sigma t)$
120	0.009	0.008	−5	0.400	0.548
100	0.020	0.019	−10	0.418	0.596
80	0.042	0.041	−15	0.432	0.641
60	0.084	0.087	−20	0.441	0.685
50	0.117	0.124	−25	0.445	0.725
45	0.138	0.146	−28.50	0.4463	0.7516
40	0.160	0.173	−30	0.446	0.763
35	0.185	0.208	−35	0.443	0.796
30	0.211	0.236	−40	0.438	0.826
25	0.240	0.273	−50	0.421	0.875
20	0.269	0.314	−60	0.399	0.912
15	0.298	0.357	−80	0.353	0.957
10	0.328	0.403	−100	0.312	0.980
5	0.355	0.451	−120	0.280	0.991
0	0.380	0.499	−150	0.245	0.997

Source: From Ref. 16, reprinted by permission.
[a]To calculate the current:
1. i = i(plane) + i(spherical correction).
2. $i = nFA\sqrt{\sigma D_O}C_O^*\sqrt{\pi}\chi(\sigma t) + nFAD_OC_O^*(1/r_O)\phi(\sigma t)$.
3. $i = 602n^{3/2}A\sqrt{D_O}vC_O^*[\sqrt{\pi}\chi(\sigma t) + 0.160(\sqrt{D_O}/r_O\sqrt{nv})\phi(\sigma t)]$ amperes at 25°.
   Units for step 3 are: A, cm^2; D_O, cm^2/sec; v, V/sec; C_O^*, moles/liter; r_O, cm.
[b]$E_{1/2} = E^{o\prime} + (RT/nF)\ln(D_a/D_O)^{1/2}$.

must be zero. Thus the differential in Eq. (1) can be approximated in a linear fashion by

$$\frac{i}{nFA} = \frac{D_o[C_o^* - C_o(r_0)]}{\delta} \tag{12}$$

This approach has been used in hydrodynamic voltammetry for many years and is known as the Nernst approximation [17]. A complete description of the voltammetric curve now requires an expression for $C_o(r_0)$ and δ, a fictitious quantity that is the thickness of the diffusion layer in the Nernst approximation (Fig. 5). If we consider the reaction $O + ne = R$ and limit ourselves to a reversible system (rapid electron-transfer kinetics, both O and R soluble in solution), the surface concentration for O [$C_o(r_0)$] is related to the applied potential (E_{app}) by the Nernst equation

$$\frac{C_o(r_0)}{C_R(r_0)} = \exp\left[\frac{nF}{RT}(E_{app} - E^{o\prime})\right] \tag{13}$$

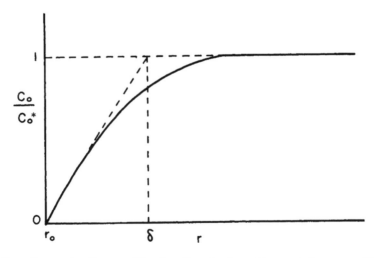

FIG. 5. Concentration profile for the electroactive species according to the Nernst diffusion-layer model for mass transport to a spherical electrode. The electrode potential is held such that the concentration of the electroactive species is zero at the electrode surface.

where $E^{o'}$ is the formal reduction potential for the couple. When E_{app} is much more negative than $E^{o'}$, $C_O(r_0) = 0$, which is the condition that we have already evaluated in the chronoamperometry case. Under steady-state conditions the current at this potential is

$$i_\ell = ar_0nFD_OC_O^*$$

(14)

where a is 2π for a hemisphere or 4π for a sphere. Thus, by substitution,

$$\frac{i_\ell}{nFA} = \frac{D_OC_O^*}{\delta} = \frac{D_OC_O^*}{r_0}$$

(15)

Equation (13) can be simplified when $D_O = D_R$, a reasonable approximation in most cases. Then $C_R(r_0) = C_O^* - C_O(r_0)$ and

$$C_O(r_0) = \frac{C_O^*}{1 + \exp[(-nF/RT)(E_{app} - E^{o'})]}$$

(16)

Substitution of the value for δ and $C_O(r_0)$ into Eq. (12) and rearrangement gives an expression for the complete voltammetric curve at a spherical electrode under steady-state conditions:

$$\frac{i}{nFA} = \frac{D_OC_O^*}{r_0[1 + \exp[(-nF/RT)(E^{o'} - E_{app})]]}$$

(17)

B. Diffusion at Disks and Rings

Fabrication of microelectrodes with the geometry of a disk embedded in an infinite insulating plane is easier to accomplish than forming spherical ultramicroelectrodes. Therefore, it is important to have analytical expressions to describe the current at electrodes of this geometry. To a first approximation one might expect the results for a disk to be very similar to those we have already examined for spherical electrodes. In particular, there should be close analogy between the results for a hemisphere and a disk, and this has been shown to be the case [5,18,19].

Thus, for small values of $D_0 t/r_0^2$, where r_0 is the radius of the disk in this case, we would expect the Cottrell equation to describe the current during chronoamperometry. For large values of $D_0 t/r_0^2$, a hemispherical diffusion layer can be envisioned over the disk, and steady-state behavior would be expected during chronoamperometry [8].

Although a conceptual feeling for diffusion processes at a disk can be arrived at without much trouble, derivation of rigorous expressions is an arduous task. The problem arises because the disk is not a uniformly accessible electrode. The flux of material reacting at the surface is unequal across the electrode surface [20] because electrolysis that occurs at the outer circumference of the disk diminishes the flux of material to the central portion of the disk [18,21]. This is not the case at spherical electrodes, which is the reason why relatively tractable solutions exist for that problem. Because of the nonuniform current distribution, the term "convergent" diffusion has been proposed to describe events at structures such as the disk. Early approaches to the mathematics of this problem used either simplifications [22] or digital simulation techniques [23] in an attempt to get a solution to the chronoamperometric problem analogous to Eq. (7). Early experimental attempts to examine steady-state behavior at disk electrodes showed that this behavior could occur, but the disks were not small enough for the predominance of this effect on the experimental time scale [24—26].

Saito was the first to give a clear insight into the steady-state nature of the current at disk electrodes [27]. He was interested in very small oxygen electrodes, and he constructed disks of several different sizes. The smallest disk had a diameter of 1.5×10^{-3} cm, and thus steady-state current could be observed. The steady-state condition means that Fick's second law can again be set to zero as was done earlier for the sphere; there is no change in concentration with respect to time in the diffusion layer. To solve the problem for a disk, a cylindrical coordinate system was used. The analogous problem has been

solved for heat conduction to a disk in an infinite insulating plane [28].
Thus Saito arrived at the correct conclusion that the limiting, steady-
state current for a disk is

$$i_\ell = 4nFD_oC_o^*r_0 \tag{18}$$

The diffusion-layer profile under steady-state conditions is given in
Fig. 6.   The steady-state solution is seen to be of a form similar to
that for a spherical electrode.   In fact, Eqs. (18) and (11) (the latter
for a hemispherical electrode) are identical when $(r_0)_{disk} = \pi/2 \times$
$(r_0)_{hemisphere}$ as originally noted by Oldham [18].   In addition, the
diffusion-layer thickness at a disk (δ) can be shown to be $\delta = (\pi/4)r_0$
through combination of Eqs. (18) and (12).

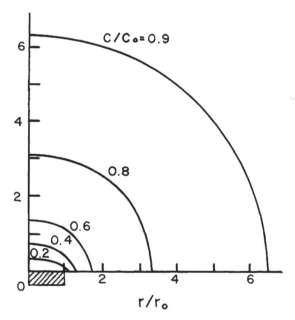

FIG. 6.   Concentration profile under steady-state conditions at a disk
electrode for times after the application of a potential step of sufficient
magnitude to drive the concentration of the electroactive species to zero
at the electrode surface.   Plotted from equations in Ref. 27.

Our initial attempts to measure these processes was with chrono-amperometry [8]. We fit the experimental data to the form seen for the hemisphere [Eq. (7)]. A plot of the measured current vs. $t^{-1/2}$ (for times from 0.1 to 10 sec) gave a reasonably straight line for a carbon fiber electrode with $r_0 = 3.5 \times 10^{-4}$ cm (Fig. 7). However, evaluation of the coefficient (a) in the empirical equation

$$i \sim nFD_oC_o^* \left[\frac{A}{(\pi D_o t)^{1/2}} + ar_0 \right] \tag{19}$$

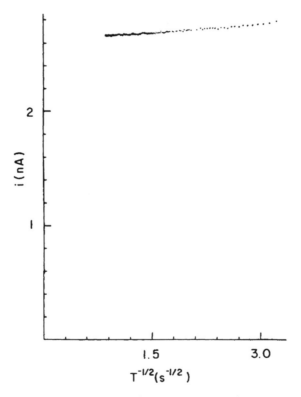

FIG. 7. Background-corrected chronoamperometric current for reduction of 4.00 mM $Fe(CN)_6^{3-}$ in aqueous 1.0 M KCl at a carbon fiber disk ultra-microelectrode ($r_0 = 3.5$ μm). (From Ref. 8, reprinted with permission.)

gave a = 5.69 ± 0.48. With a larger carbon electrode we were able to obtain a value of 3.82 ± 0.62, which is close to the value given by Eq. (18). Later, Shoup and Szabo showed that a likely reason for the high value obtained with the carbon fiber electrode is that the electrode was not embedded in an infinite insulating plane [29]. The fiber electrode used was of the "in vivo" type with the glass insulation surrounding the carbon fiber and the fiber itself had similar dimensions. At this type of electrode the steady-state term is larger than for a disk embedded in a large insulator.

The rigorous solution to the chronoamperometric problem for a finite disk was first given by Aoki and Osteryoung [30] (Shoup and Szabo made a minor correction to this result [31]). The solution is actually in two parts: a short- and long-time solution, with the current expressed as a function of a dimensionless time variable:

$$\tau = \frac{4D_o t}{r_o^2} \tag{20}$$

The current can then be expressed as

$$\frac{i}{4nFD_o AC_o^*} = f(\tau) \tag{21}$$

At short times,

$$\lim_{t \to 0} f(\tau) = 2^{-1}\pi^{1/2}\tau^{-1/2} + \frac{\pi}{4} - \frac{3\pi\tau}{2^{10}} + \cdots \tag{22}$$

whereas at long times,

$$\lim_{t \to \infty} f(\tau) = 1 + 4\pi^{-3/2}\tau^{-1/2} + 32(9^{-1} - \pi^{-2})\pi^{-3/2}\tau^{-3/2} + \cdots \tag{23}$$

Since these expressions are rather cumbersome for routine use with experimental data, Shoup and Szabo [31] have proposed the use of

$$f(\tau) = 0.7854 + 0.8862\tau^{-1/2} + 0.2146e^{-0.7823\tau^{-1/2}} \tag{24}$$

which is accurate to 0.6% for all times [32]. Kakihana et al. [33,34]
have suggested solutions of a form similar to Eq. (24), but they are not
applicable to all times. It should be noted that Oldham, using a different approach [18], arrived at a result similar to Eq. (22). Examination of the correct solution shows that the approximation given by Eq.
(19) is correct only in the fact that it has each of the first terms from
Eqs. (22) and (23).

Experimental verification of the rigorous chronoamperometric expressions has been given by Hepel and Osteryoung [35,36] for the oxidation
of ferrocyanide in 0.5 M K_2SO_4 at a gold disk ($r_0 = 1.34 \times 10^{-3}$ cm).
Similar results were obtained at carbon disks formed with photoresist
techniques [37]. Baranski et al. used the first two terms of Eq. (23)
to determine diffusion coefficients with ultramicro-disk electrodes [38].
[Note that these authors used an incorrect form of Eq. (23); since
their work relied on ratios of current, the error did not affect their
results.]

The analogy of the results at a disk and a hemisphere can also be
applied to a first approximation when cyclic voltammetry is considered.
For large values of the scan rate (small values of $D_0 t/r_0^2$), cyclic
voltammograms recorded at very small disks have the characteristics of
those recorded at a large planar electrode. With small values of the
scan rate the limiting current in cyclic voltammetry is that given by
Eq. (18). This has been experimentally verified by several groups.
For example, at carbon paste electrodes of 150-μm radius, the oxidation
of ferrocyanide is peak shaped at 0.1 V sec^{-1}, while a steady-state
voltammogram is obtained at 0.02 V sec^{-1} [19]. The limiting steady-
state current for the oxidation of ferrocene at a platinum electrodes in
acetonitrile has been shown to be a linear function of disk radius, as
expected from Eq. (18) [1].

The shape of the voltammogram under steady-state conditions for a
disk can be derived with the Nernst-layer approximation as was done
for the sphere. The current on the forward scan is given by

$$i = \frac{4nFD_oC_o^*r_0}{1 + \exp[(-nF/RT)(E^{o'} - E_{app})]} \tag{25}$$

With a carbon-disk electrode ($r_0 = 5.1 \times 10^{-4}$ cm) and the oxidation of ferrocene in acetonitrile containing 0.2 M $LiClO_4$ as a test system [5] an experimentally determined value for the coefficient (a) in Eq. (19) was determined to be 4.0 ± 0.1. In addition, it was shown that the current on the rising part of the voltammetric wave follows the expression

$$E = E_{1/2} + \frac{RT}{nF} \ln \frac{i_\ell - i}{i} \tag{26}$$

This expression, which holds only for reversible processes, is that expected for a steady-state voltammogram [39]. The current on the reverse scan is identical to that on the forward scan because the flux of zero to the electrode is independent of time and because the products formed at the electrode rapidly diffuse away from the vicinity of the electrode. Examples of cyclic voltammograms for reversible systems recorded at different scan rates are shown in Fig. 8.

For intermediate values of D_0t/r_0^2, one has to account for convergent and planar diffusion in cyclic voltammetry. To a first approximation this can be done with the hemispherical correction given by Nicholson and Shain; one simply replaces A/r_0 with $4r_0$ [5]. The peak current, normalized by $v^{1/2}$ (where v is the scan rate), is found to follow the curve over a wide range of scan rates (Fig. 9).

More complete descriptions of cyclic voltammograms can be obtained by digital simulation of the diffusion problem. Heinze has used the implicit finite-difference approach to solve problems in cyclic voltammetry [40]. The method can be used to predict the shape of voltammograms for reversible processes (Fig. 10). Pons et al. have proposed the use of simulation based on orthogonal collocation [41–44]. This method is attractive because it has inherent stability, it minimizes computational

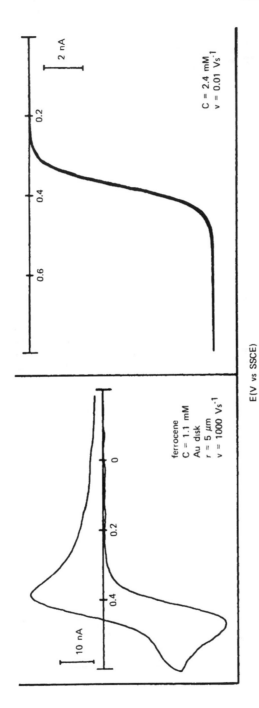

E(V vs SSCE)

FIG. 8. Cyclic voltammograms for oxidation of ferrocene in a 0.6 M TEAP/acetonitrile solution at a gold disk ultramicroelectrode (r_0 = 5.0 μm) at slow and fast scan rates.

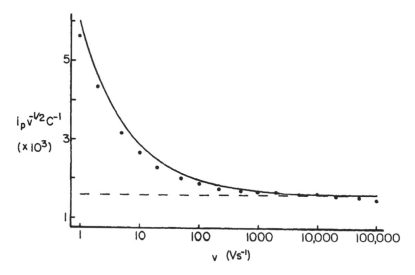

FIG. 9.   Variation of the normalized peak current vs. scan rate for a
disk ultramicroelectrode of 6.5 μm radius.   Dashed line: calculated
result for planar diffusion only.   Solid line, calculated result for both
planar and steady-state contributions; points, experimentally determined
results for reduction of anthracene in acetonitrile with 0.6 M TEAP at
a gold ultramicroelectrode.   (From Ref. 5.)

effort, and it is simple to adapt from one electrochemical mechanism to
another.   Another rapid simulation procedure is the hopscotch algorithm,
which is explicit and unconditionally stable.   This method was originally
proposed by Shoup and Szabo [45] for chronoamperometry at finite disks
but can be extended to cyclic voltammetry.   Aoki et al. [46] have
presented a numerical method to construct voltammograms for disk
electrodes.   Introducing the dimensionless parameter

$$p = \left(\frac{nFvr_0^{\,2}}{D_0 RT} \right)^{1/2} \tag{27}$$

one can compute the maximum current (i_m) in a cyclic voltammogram with
the aid of the following relation:

$$\frac{i_m}{4nFC_0^* D_0 r_0} = 0.34 \exp(-0.66p) + 0.66 - 0.13 \exp\left(\frac{-11}{p} \right) + 0.351p \tag{28}$$

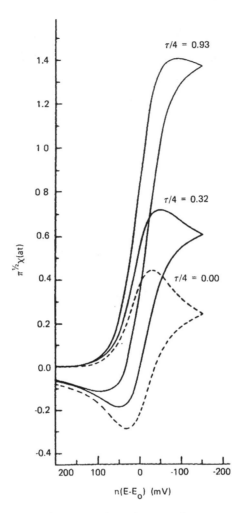

FIG. 10. Simulation of the cyclic voltammetric response at a disk
ultramicroelectrode. (From Ref. 40.)

The validity of this equation was experimentally confirmed with disk electrodes ranging in radius from 2.5×10^{-3} to 1×10^{-2} cm. In a similar way, chronopotentiometry at disk electrodes has been considered [47].

Behavior similar to that for a disk electrode is found with small ring electrodes [48]. This is because, as noted previously, most of the current at a disk arises from electrolysis at its circumference. Thus ring electrodes may be significantly more useful than disk electrodes because the part that is little used at the disk is eliminated with the ring. Theoretical descriptions of current at ring electrodes have been presented [48–50]. The limit at long times is

$$\frac{\lim_{t \to \infty} i(t)}{nFD_oC_o^*} = \ell_0\left[1 + \frac{\ell_0}{(4\pi^3Dt)^{1/2}}\right] \tag{29}$$

where ℓ_0 has the units of length and is defined for a ring of inner radius, a, and outer radius, b, as

$$\ell_0 = \frac{\pi^2(a + b)}{\ln[4e^{3/2}(b + a)/(b - a)]} \qquad \frac{b - a}{b} \ll 1$$

C. Diffusion at Cylindrical and Band Electrodes

Diffusion to and from cylindrical ultramicroelectrodes can be treated more readily than for a disk. This is because, like the spherical electrode, all points on an electrode of cylindrical geometry are uniformly accessible. Because bands and hemicylinders are geometrically similar, the currents observed at such electrodes are comparable, just as we have seen for disks and hemispheres. Therefore, although the band is probably a more useful microstructure, it is worth considering cylindrical diffusion in some detail.

For an electrode that is cylindrical and of radius r_0 and length ℓ, the solution is most conveniently obtained in cylindrical coordinates. The distance axis to be considered, r, is perpendicular to the electrode.

Contributions to the current from electrochemical reactions on the ends
of the cylinder will be ignored, which is the case when $\ell \gg r_0$. For
most experimental situations, this is a reasonable approximation. Evaluation of the diffusion equations for this situation [11] requires solution
of Fick's second law written for a cylindrical coordinate system:

$$
\frac{\partial C_o(r,t)}{\partial t} = D_o \left[\frac{\partial^2 C_o(r,t)}{\partial r^2} + \frac{(1/r)\partial C_o(r,t)}{\partial r} \right] \tag{30}
$$

The boundary conditions for chronoamperometry under diffusion-
controlled conditions are the same as those considered for spherical
diffusion. As with the disk, this problem was first solved in the treatment of heat conduction [28]. The solution is

$$
i = \frac{nFAD_o C_o^*}{r_0} f(\tau') \tag{31}
$$

where A is the area of a cylinder ($2\pi r_0 \ell$) or hemicylinder ($\pi r_0 \ell$) with

$$
\tau' = \frac{D_o t}{r_0^2} = \frac{\tau}{4} \tag{32}
$$

and

$$
f(\tau') = \frac{4}{\pi^2} \int_0^\infty \frac{\exp(-\tau' u^2) \, du}{u[J_0^2(u) + Y_0^2(u)]} \tag{33}
$$

In Eq. (33), $J_0(u)$ and $Y_0(u)$ are Bessel functions of zero order and
the first and second kinds, respectively, and u is an auxiliary variable.
The Bessel functions can be expanded [11] to give a solution for small
values of τ',

$$
\lim_{t \to 0} f(\tau') = \frac{1}{(\pi \tau')^{1/2}} + \frac{1}{2} - \frac{(\tau'/\pi)^{1/2}}{4} + \frac{\tau'}{8} - \cdots \tag{34}
$$

or a solution for large values of τ',

$$\lim_{t \to \infty} f(\tau') = \frac{2}{\ln(4\tau' e^{-2\gamma})} - \frac{2\gamma}{[\ln(4\tau' e^{-2\gamma})]^2} - \frac{\pi^2/3 - 2\gamma^2}{[\ln(4\tau' e^{-2\gamma})]^3} + \cdots$$

(35)

where γ is Euler's constant (0.57722).

The first term of Eq. (34), when substituted into Eq. (31) gives the familiar Cottrell equation, the expected answer for small values of $D_0 t/r_0^2$. This was recognized by several early workers [51–56], who showed that planar diffusion controls the events at large cylindrical electrodes used at normal electrochemical time scales perfectly well. However, with electrodes of micrometer dimensions, Eq. (35) becomes important. Delahay [11] gives the limit of Eq. (35) as

$$f(\tau') = \frac{1}{\ln[2(D_0 t)^{1/2}/r_0]}$$

(36)

Comparing this last expression to numerical values of the integral in Eq. (33), we found this expression to be accurate to within 2% for values of $\tau' > 10$ [57]. In addition, experimental data obtained at a platinum cylinder ($r_0 = 5 \times 10^{-4}$ cm, $1 = 4.85 \times 10^{-2}$ cm) are in agreement with theory (Fig. 11). Thus, unlike behavior for the disk and the sphere, the current at an electrode of cylindrical geometry is not time independent. However, as seen in Fig. 11, the approach of the current to a zero value is much slower than with planar diffusion.

For a complete description of the chronoamperometric current, it would be desirable to have a compact expression for $f(\tau')$. Aoki et al. calculated $f(\tau')$ using Simpson's 1/3 rule [58]. They were able to obtain an empirical expression which has an error of less than 1% for $0 \leq \tau' \leq 10^6$ [59]. In addition, they were able to adapt this approach to cyclic voltammetry [58], chronopotentiometry [60] and to normal and differential pulse voltammetry [61,62]. Szabo et al. have given an expression which is accurate to within 1.3% for all times [63]:

$$f(\tau') = \exp\left[\frac{-(\pi\tau')^{1/2}}{10}\right]\bigg/(\pi\tau')^{1/2} + \frac{1}{\ln[(4e^{-\gamma\tau'})^{1/2} + e^{5/3}]} \tag{37}$$

The simple form of Eq. (36) allows an estimation of the thickness of a Nernst diffusion layer at cylindrical electrodes as steady-state behavior is approached. Use of the same approach employed with spherical electrodes, one obtains

$$\delta = r_0 \ln \frac{2(D_0 t)^{1/2}}{r_0} \tag{38}$$

The exact equation for the concentration gradient for an electrolyzed species at the surface of a cylindrical electrode exists, but it is complex. It has been evaluated by Robinson et al. in spectroelectrochemical studies of the events taking place at a microcylinder electrode during chronoamperometry [64,65].

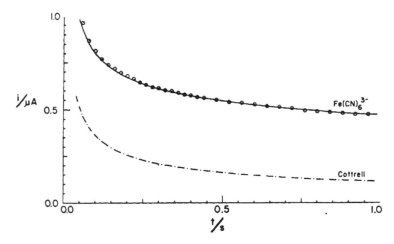

FIG. 11. Chronoamperometric results obtained with a platinum micro-cylinder electrode for reduction of 5 mm $Fe(CN)_6^{3-}$ in aqueous phosphate buffer (pH 7.4). Points, experimental data; solid line, Eqs. (31) and (36); dashed-dotted line, Cottrell equation. Experimental conditions: cylinder length = 485 μm, cylinder radius = 5.0 μm, step time = 5 sec, E_{app} = −0.2 V vs. SSCE. (From Ref. 57.)

Similar behavior for the limiting current in cyclic voltammetry is observed at cylindrical electrodes [Fig. 12(a)]. As the scan rate is decreased the shape of the voltammograms evolves from peak shaped to behavior approaching the sigmoidal curves seen at disk and spherical electrodes. The limiting current at slow scan rates (5 mV sec^{-1}) for small radii (5 × 10^{-4} cm) follows the limiting form given by Eq. (36) [57,58]. By convolution of Eq. (33) with the waveform used in cyclic voltammetry, Aoki et al. were able to generate complete voltammograms

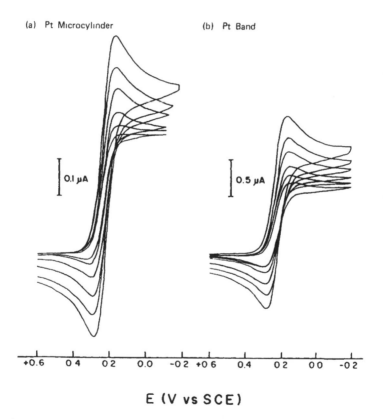

FIG. 12.   Cyclic voltammograms obtained with (a) platinum microcylinder and (b) platinum band electrodes.   Experimental conditions: scan rates, 200, 100, 50, 20, 10, and 5 mV sec^{-1} for reduction of Fe(CN)$_6^{3-}$ (5.0 mM); microcylinder length = 0.485 mm, microcylinder radius = 5 μm, band length = 0.32 cm, band width 18 μm.   (From Ref. 57.)

for reversible systems at cylindrical electrodes [58]. Cyclic voltammograms for cylindrical electrodes have also been generated with the aid
of simulation techniques [66]. To reduce the size of the space grid
that has to be considered perpendicular to the electrode surface, a
conformal map was used with the distance set to

$$y = \left(\frac{nFvr_0^2}{D_oRT}\right)^{1/2} \quad \ln\frac{r}{r_0} = p\,\ln\frac{r}{r_0} \tag{39}$$

Comparison with Eq. (38) shows that this distance coordinate is similar
to the distance of the diffusion layer at large values of y. Also with
this change of coordinate system, the gradient of concentration in the
diffusion layer becomes linear for large values of τ'. The exponential
space grid and the use of the hopscotch algorithm enable the computation time to be sufficiently short that cyclic voltammograms can be
generated within 5 min by an IBM PC computer, a considerable time
saving when compared to a finite difference approach without a conformal
map [66]. Simulated voltammograms for different conditions are given
in Fig. 13. The simulated curves are in excellent agreement with those
of Aoki et al. [58] as well as with experimental results [67].

Diffusion to band electrodes is complex because, in a fashion similar
to the events at disk electrodes, the flux is nonuniform across the
surface. In a first attempt to describe these processes, the expression
for a hemicylinder was used to approximate experimental results obtained
at band electrodes of small width ($w = 1.8 \times 10^{-3}$ cm; [57]). The
current for a hemicylinder is given by Eq. (31), where the radius used
is that for a hemicylinder of equivalent area ($r_0 = w/\pi$). As can be
seen, reasonable agreement between theory and experiment is obtained
in chronoamperometry (Fig. 14) at large values of D_ot/w^2.

Exact solution for the current in chronoamperometry at a band has
been derived by Coen et al. [68] and independently by Aoki et al. [69].

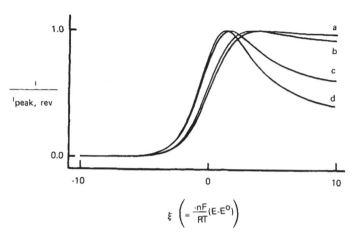

FIG. 13. Effect of the value of the cylindrical factor, β where $\beta = 2/p$. Values of β are: (a) 10^4; (b) 10^2; (c) 1; (d) 10^{-2}. For each voltammogram, $n = 1$ and $\alpha = 0.5$. (From Ref. 66.)

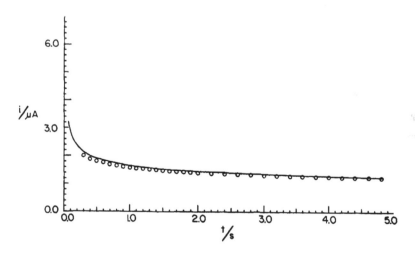

FIG. 14. Chronoamperometry at a platinum band electrode for reduction of 5 mM $Fe(CN)_6^{3-}$ in aqueous phosphate buffer (pH 7.4). Experimental conditions: step time 5 sec, $E_{app} = -0.2$ V vs. SSCE. Line, Eqs. (31) and (36) with $r_0 = w/\pi$; circles, experimental data. Electrode length = 0.32 cm, electrode width = 18 μm. (From Ref. 57.)

On the basis of these exact solutions, Szabo et al. [63] proposed empirical expressions which are accurate to within 1.3%.

$$\frac{i}{nFD_oC_o^*\ell} = \frac{1}{(\pi\tau'')^{1/2}} + 1 \qquad \tau'' < \frac{2}{5}$$

$$= \pi \frac{\exp[-(2/5)(\pi\tau'')^{1/2}]}{4(\pi\tau'')^{1/2}}$$

$$+ \frac{\pi}{\ln[(64e^{-\gamma}\tau'')^{1/2} + e^{5/3}]} \qquad \tau'' \geq \frac{2}{5} \qquad (40)$$

where

$$\tau'' = \frac{D_o t}{w^2} \qquad (41)$$

In addition, it has been shown that when the hemicylinder approximation is used at large values of $D_o t/w^2$, a better fit to the current at a band is obtained if $r_0 = w/4$ [63,69,70] rather than the approximation used by Kovach et al. It has been proposed that this approximation is exact in the limit of large values of τ''. Thus the Nernst diffusion-layer distance at a band, as the limiting current approaches steady-state behavior, will be given by Eq. (38) with the appropriate substitution for the radius.

Cyclic voltammograms obtained with band electrodes closely resemble those seen with cylindrical electrodes [Fig. 12(b)]. A quantitative way to predict the shape of these voltammograms can be obtained by means of digital simulation [71]. Because the flux across the band is nonuniform, it must be evaluated across the surface and integrated to calculate the total current. Concentration profiles generated during the simulation clearly show the hemispherical nature of the diffusion layer at large values of $D_o t/w^2$ (Fig. 15). Examination of these curves also gives a clear indication of the nonuniform values for the flux across the surface. The calculated voltammograms agree reasonably well with those obtained experimentally for the oxidation of ferrocene in acetonitrile [Fig. 16(A)]. The magnitude of the current is affected by the largest dimension of the

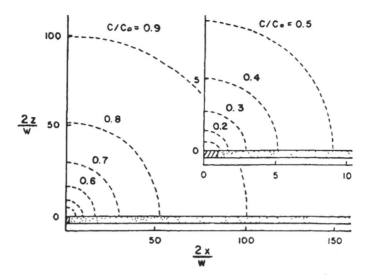

FIG. 15. Dimensionless concentration for electrolyzed species at a band electrode for conditions where the current is essentially steady state. (From Ref. 71.)

electrode, while the time dependence is affected by the width. This is in contrast to electrodes with a disk or spherical geometry, where the radius determines both the magnitude and time dependence of the current. Thus band electrodes are particularly advantageous for exploring extremely small geometries—an electrode of extremely narrow width will show quasi-steady state behavior with a very high current density, and the current can easily be measured because of the finite length of the band.

D. Diffusion at Ultramicroelectrode Arrays

A common feature of all the microelectrode geometries is that as steady-state behavior is approached, the dimensions of the diffusion layer are of finite size. As has been seen, the Nernst diffusion layer has dimensions proportional to the smallest dimension of the electrode. Thus, if two or more microelectrodes are operated at the same potential, the total response of this array of electrodes will depend on the spacing

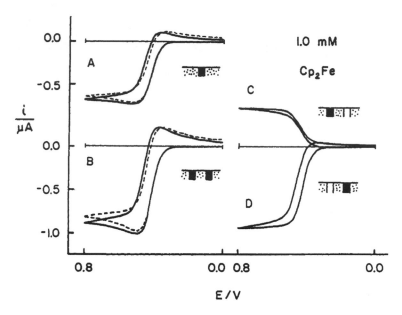

FIG. 16.   Oxidation of ferrocene (1 mM) in 0.1 M tetrabutylammonium
hexafluorophosphate/CH_3CN at single- and paired-band electrodes.
(Width of bands: 4.6 μm; scan rate: 0.1 V sec^{-1}).   (A) and (B) Solid
lines, oxidation of ferrocene at a single-band electrode (A) and at
adjacent bands separated by 4 μm (B); dashed lines, simulation for
voltammograms at 4.6-μm and at 13-μm bands, the width of the two bands
plus the insulator.   (C) and (D) Response of collector and generator
electrodes, respectively, when the potential of the collector is held at
0.0 V and the generator is scanned.

between the individual elements of the array.   If the individual elements
are spaced sufficiently far apart, the total response will be the sum of
the responses of the individual electrodes [37,72].   However, if the
spacing is sufficiently small such that the diffusion layers overlap as
steady state is approached, the total response will be less than that
predicted by the sum for each.

   To describe the current under these conditions requires consideration of the experimental time as well as interelectrode spacing.   At very
short times the diffusion layer at each ultramicroelectrode acts independently, but as time elapses, the individual diffusion layers grow and

start to interact. At sufficiently long times and with sufficiently close spacing between the individual elements, the array will approach the behavior of a large electrode whose area is given by the sum of the individual electrode areas and the intervening insulator. In fact, the original consideration of this type of problem was by investigators examining the effects of blockage of regions on a large electrode [73–76]. Gueshi et al. were able to fabricate partially blocked electrodes consisting of regularly spaced disks with photoresist techniques. They were able to show that the observed voltammograms appeared to be quasireversible under these conditions, although the actual rate of electron transfer was rapid. Amatore et al. have considered a similar problem [77,78] and have shown its importance in measurements of electrochemical kinetics by means of cyclic voltammetry. Digital simulation of cyclic voltammograms at an array disks has been presented [79].

The most complete set of expressions to describe chronoamperometry with an array of disks is given by Shoup and Szabo [80]. The work is based on earlier results provided by Reller et al. [81,82], who used explicit, finite-difference simulations. The array is considered to be an hexagonal arrangement of disk electrodes, each of radius r_0. The hexagonal insulator surrounding each disk is approximated by a disk of radius R. Thus the fraction of the surface that is inert is

$$\theta = 1 - \left(\frac{r_0}{R}\right)^2 \tag{42}$$

The problem which is formulated with the aid of a cylindrical coordinate system (as is done with a single disk) requires solution of the expression

$$i = 4r_0 nFD_o C_o^* f(\tau, \theta) \tag{43}$$

where τ is defined by Eq. (20) and the current is given for a single unit cell of the array. When θ approaches 1, the solution is given by that for an isolated disk. This solution is appropriate for

$$(R - r_0)^2 > 6D_0 t \qquad\qquad\qquad\qquad (44)$$

The response at intermediate values of the space and time parameters requires simulation (Fig. 17), and the hopscotch method has been used [80]. A transition can be noted in these curves at the time where the diffusion layers overlap, and the current decreases at each individual electrode. This process is referred to as "shielding," since electrolysis at adjacent electrodes shields individual elements from the flux they would normally experience.

Diffusion to arrays of band electrodes has been evaluated [83,84] via digital simulation. Closely spaced bands operated at the same

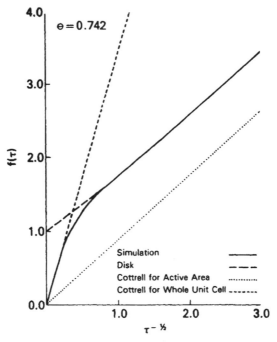

FIG. 17. Simulated dimensionless current to an ensemble of microdisk electrodes compared with the isolated disk result and the Cottrell currents to the active area and to the entire area of the unit cell. (From Ref. 80.)

potential also tend to shield one another. This is shown in Fig. 16(B), where the current at a single band for the oxidation of ferrocene is compared to that obtained when an adjacent band is operated at the same potential. Simulated voltammograms are superimposed on both measured traces. For the simulation of the paired band, the width was taken as the sum of the widths of the individual elements as well as the insulator. The agreement of the experiment with the simulation indicates one of the fundamental parameters of diffusion-controlled electrochemistry at bands that has already been discussed—most of the flux is at the outer edges of the band. With the paired bands, the inner edges are shielded and thus the response is more like a much wider electrode. The fact that electrochemical reactions do not occur in the center when an insulator is present is thus a close analogy to what actually occurs at a solid band.

The amperometric response of band electrodes has also been addressed under conditions of convective mass transport [85–90]. When the direction of flow is perpendicular to the major axis of the bands, a larger current density is obtained relative to a solid electrode in a channel of the same dimensions as a band. This is because the insulator between the bands provides a region for the diffusion layer to refill before the next band is encountered. This feature has practical implications for electroanalysis in flowing streams such as liquid chromatographic detectors.

Arrays of bands have also been operated with elements at different potentials. In this way an adjacent band (termed the collector) can be used to sample the products of the primary electrolysis at a generator electrode [83]. This phenomenon is also indicated in Fig. 16 for the paired band electrode using ferrocene as an example. In this experiment, a cyclic voltammetric experiment is run in a conventional fashion at the generator. The potential of the collector is held at the initial potential and the current is recorded as a function of the generator potential. The current is of opposite sign because, in this example, the ferrocenium cation produced at the generator electrode is reduced at the collector electrode. Thus the results at the paired band are similar to those obtained with a rotated ring-disk electrode (RRDE), the fraction collected (or collection efficiency) being dependent on the spacer thickness [83]. Unlike the RRDE experiment, the current at the generator electrode is larger than the current observed when operated independently. This is because ferrocene is regenerated by the collector electrode, so there is a greater flux of ferrocene to the generator electrode, a process referred to as feedback.

The collection efficiency for a triple-band electrode with a central generator and two flanking collectors has been evaluated as a function of gap width by means of digital simulation. For a reversible couple the collection efficiency exceeds 0.8 for gap width of 1×10^{-4} cm [83]. In addition, this response has been shown to be relatively insensitive to the width of the collector electrodes.

E. Summary

Thus it can be seen that in the past decade, considerable progress has been made in the understanding of diffusion processes to ultramicroelectrodes. Solutions exist for the current at many different microstructures, including the sphere, disk, cylinder, band, and ring. Predictions can also be made of the current at arrays of these geometries. Recently, it has been noted by Szabo [50] that the solution for the chronoamperometric current at large values of $D_0 t / r_0^2$ can be placed in a form given by Eq. (29) for several geometries of ultramicroelectrodes having a closed form. If this is the case, this provides a simple way to predict the way in which the current approaches a steady-state level. The steady-state result is much simpler to derive than the time-dependent one, and once it is known, the appropriate form of ℓ_0 can be derived for the particular geometry and employed in Eq. (29). These results

are important because they allow predictions to be made of the perform-
ance of different geometries of ultramicroelectrodes for a variety of
applications, including analytical applications and mechanistic investigations.

III. ADVANTAGES OF ULTRAMICROELECTRODES

To understand the advantages that occur with very small electrodes, we
first need to consider the flow of current through an electrochemical
cell. For simplicity, consider a two-electrode system that contains a
large, low-impedance reference electrode. The cell contains a solution
including an inert, dissociated salt. The solution has a finite resistance
(R), characterized by the specific resistivity (ρ) which has units of
Ω-cm. The working electrode, made of an inert material, has an associated capacitance that arises because charge cannot cross the solution-
electrode interface unless faradaic electrochemistry occurs. The apparent
capacitance (C) arises from the structure of ions in the double layer,
and models exist to describe this structure at a molecular level [9].
The value of the double-layer capacitance is on the order of 20 μF cm^{-2}
for most common electrode materials in aqueous solution.

Thus, in the absence of faradaic events, the electrochemical cell has
electrical characteristics similar to a resistor-capacitor series network
[9]. When the potential of the electrode is changed, the impedance of
the cell responds to this change in a similar way to that of the equivalent circuit. If a potential step ΔE is applied to an RC network, the
current will flow as the capacitor is charged or discharged. This
current, referred to as charging current (i_c), is given by

$$i_c = \frac{\Delta E}{R} \exp\left(\frac{-t}{RC}\right) \tag{45}$$

If the potential applied to the RC network is a triangular wave with
slope v, then

$$i_c = vC + \left(\frac{E_i}{R} - vC\right) \exp\left(\frac{-t}{RC}\right) \tag{46}$$

where E_i is the initial potential. In most cases the second term is insignificant, so

$$i_c = vC \tag{47}$$

If a faradaic species is added to the cell, the total current is the sum of the charging current and the faradaic current, which is given by the appropriate expression of Sec. II.

A. Effects of Reduced Capacitance

The charging current can be a major interference since it may exceed the value of the faradiac current at short times or with low concentrations of electroactive species. However, since the magnitude of the double-layer capacitance is proportional to the electrode area, it is much reduced at ultramicroelectrodes relative to electrodes of conventional size. When an ultramicroelectrode is operated under conditions where the faradaic current is steady state, this can lead to considerable improvement in the faradaic-to-charging current ratio (i_F/i_c). For example, at a disk or a spherical electrode the steady-state current is proportional to the radius, rather than to the area of the electrode, and the i_F/i_c ratio will improve inversely with the value of the radius. The improvement is not as dramatic with cylinder electrodes [i.e., Eqs. (31) and (36) combined], but appreciable improvements are still obtained by decreasing the electrode radius. This was demonstrated at a band electrode for the reduction of ruthenium (III) hexaammine at platinum electrodes in aqueous solution [91]. The i_F/i_c ratio was increased 100-fold at a band electrode (30-nm width) relative to a disk at 2-nm diameter when both were used at 5 mV sec^{-1}. For the same reasons, improved signal-to-noise ratios are observed in photoelectrochemical experiments at carbon fibers relative to electrodes of conventional size [92].

The traditional strategy to improve i_F/i_c ratios under conditions of linear diffusion has been to use pulsed voltammetric schemes. The charging current decays exponentially with time, while the decay in faradiac current is proportional to $t^{-1/2}$ under conditions of linear diffusion. As noted in Sec. II, the decay in faradaic current is much less time dependent at ultramicroelectrodes, especially at large values of $D_0 t/r_0^2$. Thus pulsed voltammetric techniques should be even more successful with ultramicroelectrodes if the background current is purely capacitive in nature. This has been shown to be the case. For example, micromolar levels of catecholamines can be detected with normal pulse voltammetry at carbon fiber electrodes [93]. The signal can be further improved by adding the current from the step back to the initial potential to correct for the residual current on the initial potential step. Very little faradaic current occurs on the reverse step of a chrono-amperometric experiment at an ultramicroelectrode because convergent diffusion allows the electrolysis products to leave the vicinity of the electrode rapidly. Therefore, the reverse current can be used to remove almost completely other sources of current from the faradaic current. In addition, much lower concentrations of catecholamines can be determined with pulse voltammetric techniques [94,95] if the carbon surfaces are electrochemically pretreated to adsorb these substances [67,96].

Square-wave voltammetry has also been examined at ultramicroelectrodes, and an exact theory has been developed for very small disks [97]. It has been shown that the shape of the net current, square-wave voltammogram is independent of the electrode geometry [98]. This means that the peak potential corresponds to the half-wave potential, irrespective of the geometry of the electrode. Thus it has been proposed that the use of square-wave voltammetry with microelectrodes is especially useful for voltammetric measurements because the shape of the signal depends primarily on the properties of the electrolyzed species, and the precise geometry of the working electrode does not require complete definition.

Trace detection of ferrocene in acetonitrile with microdisks and square-wave voltammetry has been demonstrated [99]. In this example, synchronous demodulation of the current with a lock-in amplifier was used to improve signal-to-noise ratios. The charging current was removed from the modulated current by an electronic switch. The combined use of these two methodologies led to a detection limit of 2×10^{-7} M at a modulation frequency of 200 Hz. More dramatic detection limits were obtained with hydrodynamic modulation achieved by vibrating a microcylinder. In this case, 3×10^{-8} M ferrocene in acetonitrile could be detected [100]. High frequencies for the hydrodynamic modulation are possible because of the very small dimensions of the electrode. Since the faradaic signal is the primary one that is modulated, signal-to-noise ratios are improved. The use of ac voltammetry can also be used to distinguish between faradaic and charging current, and this approach has been investigated by Baranski at microdisks [101]. The features of pulse techniques at band electrodes have been demonstrated [84,102].

B. Effects of Solution Resistance

In addition to the reduced effects of double-layer capacitance at ultramicroelectrodes, the effects of solution resistance are reduced as well. Solution resistance can severely distort electrochemical data, and until the advent of ultramicroelectrodes, precluded measurements in many solvents. This is because the current that flows through solution, the sum of the faradaic and charging currents, generates a potential that opposes the applied potential. In other words, the ohmic drop (given by the product iR) will be subtracted from the applied potential difference between the working and reference electrodes. At electrodes of conventional size the problem is circumvented to a certain degree by the use of a three-electrode potentiostat. In this design, current flows between the working electrode and a counter or auxiliary electrode. A

reference electrode is placed very close to the working electrode, where
it senses the potential difference. The iR drop is then partially compensated by electronic means. Further compensation can be achieved
by electronic positive feedback. However, this can result in oscillation
of the electronic circuit [103].

The magnitude of the resistance (R) that affects the current can be
expressed in a quantitative way for many different electrode geometries.
One model that has been used to represent the resistance at a spherical
electrode is to consider the case of two concentric spheres [104,105].
Conduction is through the medium between the spheres and the medium
is considered isotropic and with a specific resistivity ρ. For the
spherical case the solution is

$$R = \rho \left(1 - \frac{r_0}{d} \right) (4 \pi r_0)^{-1} \tag{48}$$

where r_0 is the radius of the inner sphere and d is the distance between
the spheres. To maximize the ratio r_0/d, the reference electrode should
be placed as close as possible to the working electrode. However, this
is mechanically difficult with an ultramicroelectrode and the resistance
under normal operating conditions ($r_0 \ll d$) is

$$R = \frac{\rho}{4 \pi r_0} \tag{49}$$

In a similar maner, an expression for the resistance at a microdisk with
a remote reference electrode [106] can be derived:

$$R = \frac{\rho}{4r_0} \tag{50}$$

The resistance between two concentric cylinders in an isotropic medium
is given by

$$R = \frac{\rho}{2 \pi \ell} \ln \frac{r}{r_0} \tag{51}$$

where r is the radius of the larger cylinder [107]. A more complex expression for the cylinder can be written which also accounts for the internal resistance of the cylinder [65]. In addition, Eq. (51) has been proposed to be useful for electrodes with the geometry of a band [91, 107].

To estimate the ohmic drop, these equations for resistance must be combined with the appropriate faradaic and charging current expressions. The ohmic drop caused by charging current can be diminished with a decrease in electrode area. This is because the charging current is directly proportional to the area, while each of the expressions for resistance is inversely proportional to the smallest electrode dimension. A similar improvement is seen for the ohmic drop which results from the faradaic current under conditions of planar diffusion, since the current is again proportional to the electrode area. Evaluation of the ohmic drop at a sphere or a disk under conditions of steady-state faradaic current suggests that the ohmic drop cannot be further reduced by a decrease in the smallest electrode dimension [5]. In fact, it has been shown that the ohmic drop should be independent of the geometry of the electrode under conditions where true steady-state currents are obtained [108].

The predictions based on these elementary considerations of theory have been experimentally tested [4]. For example, at a disk electrode of radius 6.5×10^{-4} cm, the voltammograms are virtually undistorted by ohmic drop when ferrocene is oxidized in acetonitrile with a concentration of supporting electrolyte one-tenth that of the depolarizer (Fig. 18). The specific resistivity of this solution is 600 kΩ-cm. Furthermore, the distortion of the voltammetric waves is relatively small even with no deliberately added electrolyte [109]. Trace impurities in the solution may serve as the solution conductive species in this case. Nevertheless, as pointed out by Fleischmann, electrochemistry without supporting electrolyte is of fundamental interest because it could lead to situations where a single ion occurs [110].

In these experiments with microdisks, the currents are in the nano-ampere range. With a disk of conventional size (r_0 = 2 mm), at which

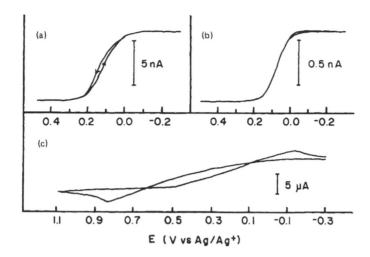

FIG. 18. Cyclic voltammograms for oxidation of ferrocene in acetonitrile with various supporting electrolyte concentrations. (a) 1.1 mM ferrocene with 0.01 mM TBAP at a 6.5-μm-radius gold microdisk electrode; arrows indicate scan direction. (b) As (a) but with 0.11 mM ferrocene. (c) As (a) but at 0.4-mm-radius platinum disk electrode.

the current shows the behavior expected for planar diffusion and is in the microampere range, the voltammograms are distorted by ohmic drop to such an extent that they are unrecognizable even with the use of a Luggin probe and a three-electrode potentiostat (Fig. 18). However, it is not the amplitude of the current that causes the ohmic distortion but its distribution through solution in the vicinity of the electrode surface. For example, with microband electrodes, which produce quasi-steady-state currents of microampere amplitude, nondistorted voltammograms are obtained for the oxidation of ferrocene under identical conditions even though large currents flow [84,102].

Careful inspection of the published voltammetric data indicates that the observed ohmic drop is actually less than predicted by the combination of the equations for resistance and the faradaic current [5]. In contrast to the results predicted above, the effects of ohmic drop tend to be even less with disk electrodes of smaller (5 × 10^{-5} cm) radii [110].

There are several reasons why this may be the case. In the case of ferrocene oxidation, the product is a charged species, the ferricenium cation. The concentration of the cation is highest in the diffusion layer, and this is exactly the region of highest resistance, because the lines of current are most densely spaced in this region. Thus one generates a conductive medium as the voltammetry occurs. Evidence for this can be seen from the hysteresis on the reverse scan of voltammograms in resistive media (Fig. 18). The generation or consumption of charged species by the electrochemical media has important consequences for the mass transport mode as well. When the electrode participants are involved in the transport process, migration as well as diffusion must be considered [110,111]. Another factor that affects the ohmic drop at disk and band electrodes is the nonuniform current distribution that normally exists at these electrodes [110]. The ohmic drop is greatest at the perimeter, and this will cause a greater flux to the usually shielded inner surface of the electrode. Thus, for several reasons, the equations presented earlier for resistance tend to overestimate the measured effects.

The ability to use electrochemistry in highly resistive solutions has inspired a number of explorations into solvents in which electrochemistry is usually not conducted. Lines and Parker were the first to show that voltammetry at ultramicroelectrodes is possible in benzene containing tetrahexylammonium perchlorate (THAP) [112]. They examined the reduction of perylene with an electrode 30 μm in diameter. The voltammetric data were difficult to interpret because intermediate values of D_0t/r_0^2 were employed. With the use of larger values of the τ, Howell [113] was able to show that undistorted voltammograms for ferrocene could be obtained in solutions containing THAP (0.5 M) even in solvents with a dielectric constant as low as 5.6 [distortion was evaluated with the use of Eq. (26)]. In solutions of benzene or toluene, the effect of ohmic drop for the oxidation of ferrocene is apparent at ultramicroelectrodes [113–116], but the problem is less severe with disks of submicrometer dimensions [116]. The voltammetric behavior is complicated on the reverse scan by a wave for the oxidation product. As noted, a reverse wave is normally absent because of rapid diffusion away from the electrode. The wave seen in toluene, benzene, and hexane has been attributed to the insolubility of the charged product, which appears to deposit on the electrode surface [116,117] in these low-dielectric solvents.

For work in solutions of low dielectric constant, the selection of supporting electrolyte is critical—the salt used must be soluble and dissociate to some degree. This has been investigated in some detail by Murray's group [114,115]. With the use of viscosity measurements, vapor osmometry, and impedance measurements, it has been proposed that the electrolyte, THAP, actually aggregates to trimers and tretramers in toluene. Nevertheless, sufficient dissociation occurs that voltammetry is possible. Undistorted voltammograms for ferrocene have been achieved in CO_2 under supercritical conditions with added water and tetrahexyl-ammonium hexafluorophosphate (Fig. 19). Presumably, a similar dissociative mechanism occurs in this solution. Quatenary ammonium salts are not soluble in heptane. However, the addition of mixtures of fatty acids and bases renders this solvent sufficiently conductive for voltammetric measurements [114].

Although problems with ohmic drop are minimized with ultramicro-electrodes, the voltammetric data for resistive media reflect a number of factors that are normally not concerned with electrochemistry in conventional solutions. For example, the selection of a reference potential needs to be established, especially if results are to be compared between solutions. Conventional, aqueous-based reference electrodes are unsuitable because of problems with contamination of the solvent, miscibility of the solutions, and the likelihood of large liquid-junction potentials. A preferable approach is to use a quasi-reference electrode, such as a silver wire, and then refer all potentials to a standard test couple. A relative potential scale has been published which compares half-wave

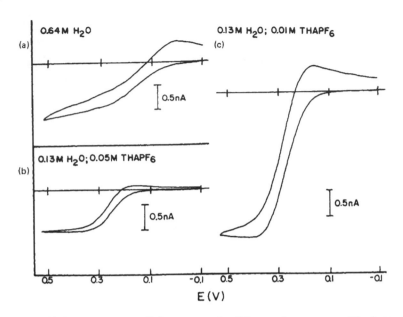

FIG. 19. Voltammograms of ferrocene in CO_2 under supercritical conditions (80°C and 88 atm) at a microdisk electrode (r_0 = 5 μm). Potentials were recorded against a silver wire quasi-reference electrode; the scan rate was 100 mV sec^{-1}. (a) Oxidation of 19 μM ferrocene in CO_2 containing 0.64 M H_2O. (b) and (c) Oxidation of 47 μM ferrocene in CO_2 containing 0.13 M H_2O and 0.05 or 0.01 M tetrahexylammonium hexafluorophosphate, respectively. (From Ref. 118.)

potentials of several compounds in toluene and acetonitrile. Strangely, the shift for ferrocene differs greatly from other one-electron couples [114]. Ferrocene has traditionally been used as a reference, but the published data suggest that other couples may be more useful.

Another factor, alluded to earlier, is that migration may play a role in the mass transport of the electroactive species. Migration occurs when an electric field exists in solution. This is the case in the diffusion layer in any amperometric experiment because of the generation or consumption of ions by the electrode. However, normally the migrating species is a purposely added, inert electrolyte, and thus the magnitude of the faradaic current is not affected. Under conditions where the concentration of the electroactive species exceeds that of the inert electrolyte, the mass transport of at least one of the partners of the redox reaction will be affected by the migration process. For example, the limiting current for the oxidation of the neutral species, ferrocene, is unaffected by migration at low supporting electrolyte concentrations, although calculations show that the concentration of the electrogenerated cation will be lower in the diffusion layer when migration occurs as compared to diffusional control [111]. In contrast, the limiting current for the reduction of cobaltecinium is found to be significantly increased relative to that expected under diffusion control with a low concentration of inert electrolyte. In this case there is an increased flux of the cation to the electrode because of the generation of a neutral species by the electrolysis and the resulting need to restore charge balance. In addition to the effects of migration, diffusion coefficients alter with changes in supporting electrolyte concentration. This is because the inert salt increases the viscosity of the solution.

A third complication of working with a low supporting electrolyte concentration is that the distance that the double layer extends into solution will be larger than in conventional experiments. This may result in large effects on electrode kinetics because the rate constant depends on the potential difference between the electrode surface and the plane of closest approach of the molecule to the electrode [9]. Changes in the apparent kinetics are not present in the published experimental data; however, this aspect has not yet been sufficiently well explored.

As well as exploring solvents of low dielectric constants, other resistive media have been explored. Among the first reported was the voltammetry of ferrocene at or near the glass point of acetonitrile [119]. Surprisingly, there was little change reported in the magnitude of the limiting current at the glass point, suggesting that mass-transport processes are unchanged. Voltammograms have been recorded at potentials beyond that where the solvent is electrolyzed. For example,

Malmsten et al. recorded voltammograms of pure nitrobenzene which contained 0.1 M tetrabutylammonium perchlorate [120]. The data from a microdisk clearly show a limiting current for the reduction of the solvent (Fig. 20). The wave shape for the first wave is that expected for a one-electron reduction of nitrobenzene. At more negative potentials a wave attributed to the formation of the dianion is observed; however, its amplitude is greatly attenuated. Pons's group has reported waves for the oxidation of alkanes [121] and various gases [122] beyond the potential limit of acetonitrile. In these studies the ring geometry was employed and supporting electrolyte was not used. It was found that the oxidation of the solvent formed a polymeric film on the electrode. Nevertheless, the film was permeable to small molecules, and limiting currents that depend on concentration could be obtained. Other media that have been explored include microemulsions [123] and aqueous solutions with little or no electrolyte [124]. Electrochemistry has even been observed in the absence of solution [7] with an electrode comprised of two concentric rings. In this case it is presumed that the conductive phase is the epoxy between the two rings. The device was shown to be suitable as a gas chromatographic detector for a wide number of substances.

C. Fast Electrochemistry

The capacitive nature of the impedance of any electrochemical cell precludes rapid changes in potential. Thus electrochemical measurements with conventional electrodes have generally been restricted to the millisecond, or longer, time scale. Various schemes have been devised to improve the time resolution of electrochemical measurements, but these have been instrumentally complex [125—128]. However, the ability to make fast electrochemical measurements is desirable so that rapid heterogenous or homogeneous reaction rates can be measured.

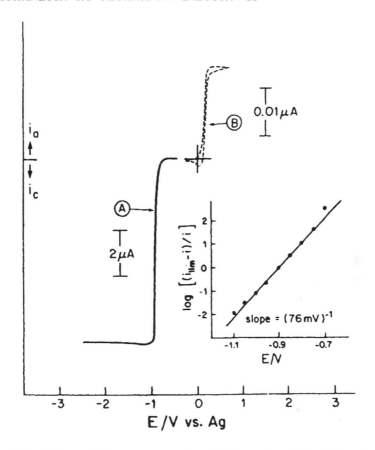

FIG. 20. Voltammetric response at a platinum microdisk electrode (r_0 = 12.5 µm) in: (a) nitrobenzene containing 0.1 M TBAP; and (b) the same solution containing 10 mM decamethylferrocene. Scan rate 20 mV/sec. Note change in current scale. Inset: Plot of log $[(i_{lim} - i/i]$ vs. V corresponding to curve a. (From Ref. 120.)

McCreery's group was the first to demonstrate that the reduced values of resistance and capacitance found at ultramicroelectrodes lower the cell-time constant, and thus allow the working electrode to assume the applied potential at a very short time scale [64]. Rather than monitor the current, they used absorption spectroelectrochemistry to measure the chemical events at the electrode surface. In this way they avoided complications associated with large charging currents. The

absorption, measured in an external reflection mode, has been used to
monitor the oxidation of o-dianisidine on a microsecond time scale at a
cylindrical electrode [65]. With the use of a microdisk electrode (r_0 =
3×10^{-3} cm) the time scale can be extended into the nanosecond time
domain [129]. The improvement in time resolution with the disk was
made possible because it has a lower time constant than the cylinder.

Electrochemistry at these very fast time scales follows planar diffusion (i.e., $D_0 t / r_0^2 < 1$) at ultramicroelectrodes. Therefore, the voltammetric current, as well as the charging current, are both proportional
to the electrode area. For this reason the i_F / i_c ratio is exactly the
same as would be expected to occur at large electrodes, and the current
becomes uniformly distributed across the surface. The advantages with
the ultramicroelectrode under these conditions are the low values of the
currents, which reduce ohmic drop, as well as the low cell-time constant.
However, the faradaic current during cyclic voltammetry, which increases
with $v^{1/2}$ for planar diffusion, may be obscured by the charging current,
which increases linearly with v [Eq. (47)]. Nevertheless, voltammograms
with useful voltammetric information can be obtained at scan rates as
high as 10^5 V sec^{-1} with disks of micrometer dimensions [5] (Fig. 21).
Furthermore, these fast scan rates can be used in solutions of relatively
high resistance [113].

Fast-scan cyclic voltammetry has been used to study several chemical
systems. The fast-scan technique is particularly useful to outrun
chemical reactions which consume electrogenerated products. For example,
the dication of diphenylanthracene is stable only on a microsecond time
scale in acetonitrile [113]. In another example, it has been shown that
the radical cations of several aromatic hydrocarbons are stable on this
time scale. Therefore, the E^{o}'s could be determined, and these were
correlated with their vertical ionization potentials [130]. The disproportionation of quinone radical anions at basic pH has also been investigated
[131]. The oxidized form of ascorbic acid, which has a half-life of less

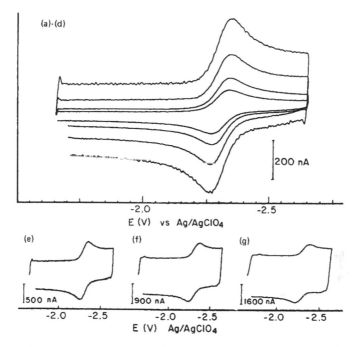

FIG. 21. Cyclic voltammograms for reduction of anthracene (2.22 mM) in acetonitrile with 0.6 M TEAP at a gold microdisk electrode (r_0 = 6.5 µm): Scan rates in V sec^{-1}. (a) 1000; (b) 2000; (c) 5000; (d) 10,000; (e) 20,000; (f) 50,000; (g) 100,000. (From Ref. 5.)

than a millisecond, can be observed with this technique (Fig. 22). Similarly, structural changes of the reduced form of bianthrone, which occur on a microsecond time scale, have been determined [132]. The use of fast-scan voltammetry at ultramicroelectrodes has also been shown to enable the measurement of heterogeneous electron-transfer reactions with rate constants exceeding 2 cm sec^{-1} [5,133].

Thus it is clear that ultramicroelectrodes facilitate very fast measurements. However, these measurements are only possible with properly constructed electrodes. If solution leaks into the space between the insulator and the electrode material, the apparent capacitance will be much larger than predicted, and the faradaic current will be obscured by the double-layer processes [134]. A development that facilitates

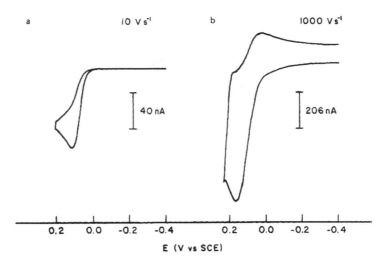

FIG. 22. Cyclic voltammogram for oxidation of ascorbic acid (6.8 mM) in 0.1 M phosphate buffer (pH 7.0) at a mercury ultramicroelectrode (r_0 = 5.0 μm). (a) 10 V sec^{-1}; (b) 1000 V sec^{-1}. (From Ref. 15.)

trace analysis with fast-scan voltammetry [135] as well as enabling much faster scan rates to be observed, is the use of digital subtraction of the capacitive or background current [136]. As long as the background features do not change, the subtraction leads to undistorted voltammograms (Fig. 23). This is the case when the electrode is used in a flow-injection-analysis system.

D. Applications that Exploit the Steady-State Current

The steady-state nature of the current obtained at large values of D_0t/r^2 can alter the way in which chemical reactions that accompany electron transfer manifest themselves in the measured current. For reactions that follow the initial electron transfer, the reactions tend to be less apparent. For example, in an ECE reaction, the product formed during the intervening chemical reaction is less likely to return to the electrode to undergo a subsequent electron transfer. This is true if the chemical reaction has a lifetime which is sufficiently long that the chemical product has time to diffuse away from the electrode. Since the

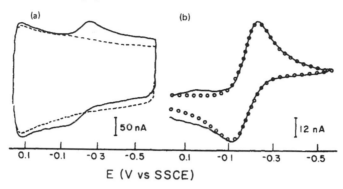

E (V vs SSCE)

FIG. 23. Cyclic voltammograms for reduction of 1 mM ruthenium(III) hexaammine in an aqueous solution of 0.1 M sodium trifluoroacetate at a hemispherical mercury ultramicroelectrode (r_0 = 5.5 μm) at a scan rate of 2000 V sec^{-1}. (a) Dashed line, background current; solid line, background plus faradaic current. (b) Solid line, subtracted cyclic voltammogram; open circles, simulation of cyclic voltammogram for α = 0.7 and k° = 0.52 cm sec^{-1}.

dimensions of the diffusion layer tend to be very small at ultramicroelectrodes, this can easily occur. This has been demonstrated for the intracyclization of dopamine-o-quinone, an ECE reaction, where the second electron transfer is much less apparent than at an electrode where planar diffusion is predominant [8]. The effects of catalytic reactions are also less apparent. For example, the catalytic oxidation of ascorbate by the o-quinone of dopamine is much less apparent at carbon-microdisk electrodes operated under steady-state conditions than at larger electrodes [137]. This result has important implications for electroanalysis since the measured current is less sensitive to secondary processes that occur in the diffusion layer.

For many cases in which chemical reactions are coupled to electrochemical reactions, expressions exist for chronoamperometric current at spherical electrodes [17]. These can be used to predict the degree to which these reactions will manifest themselves. These expressions show

that the measured current will be a function of the chemical rate constant
and the electrode radius for steady-state conditions. The expressions
have been adapted to determine rates and mechanisms at microdisks with
considerable success [137–139]. The catalytic effect of aminopyrine on
the oxidation of ferrocyanide has been examined as a function of electrode
radius [138]. The rate constant evaluated from the experimental data
is in excellent agreement with that measured by conventional electrode
methodology. For the reduction of protons in buffered solutions of
acetic acid, a rate constant for the dissociation reaction near the diffusion limit was mesured with this approach [138]. Fleischman and co-
workers have also shown that the rate constants for ECE reactions can
be determined in this way [139]. In a study of the oxidation of anthracene in acetonitrile, a one-electron oxidation reaction followed by a
rapid CE reaction, they showed that the limiting one-electron behavior
can be obtained with very small electrodes. In addition, they concluded
that the overall reaction was a true ECE sequence, as opposed to a
disproportionation reaction, based on agreement of the measured rate
with that obtained at a rotated disk electrode. Digital simulation procedures can also be used to predict the effect of accompanying chemical
on the steady-state response [140].

Because the current density is very high at ultramicroelectrodes,
the effects of finite rates of electron transfer are much more apparent
under steady-state conditions. An early kinetic study that took advantage of this feature employed an ensemble of 10^6 microelectrodes (in the
form of mercury droplets) with an average radii of 100 nm [141,142].
The rising part of the voltammetric wave has been used to determine
electron-transfer rates at ultramicrocarbon electrodes [137]. However,
because of the nonuniform current distribution at disks, these measurements give only an approximate measure of the rate constant. A more
suitable electrode is the ring electrode because the current density is
more uniform [143].

Ultramicroelectrodes, operated under conditions where steady-state behavior occurs, tend to be relatively immune to convection. This is because the flux as a result of diffusion is very high, and convection adds little mass transport [72] except at high rates of stirring [100]. Because of this, arrays of microelectrodes have been used as detectors in liquid chromatography [72,144—146]. An advantage in signal-to-noise ratio has been shown over electrodes of conventional size in circumstances where flow rate fluctuations caused by the pump are predominant [72]. This is most likely to be the case at extreme potentials where background currents are high. Another advantage obtained with arrays of microelectrodes as liquid chromatographic detectors is the high ratio of faradaic to charging current found under steady-state conditions. The combined effect of both of these features may explain why composite electrodes, such as carbon paste, have superior detection limits to solid electrodes as liquid chromatography detectors. The conductive particles in composites may act as microelectrodes separated from one another by the binding material [147]. Microelectrodes are particularly well suited for detection following separation by normal-phase chromatography where nonpolar mobile phases are used. This is because they can be used in resistive media without the problems associated with ohmic drop [144,145].

As noted earlier, at very high convection rates the current at microelectrodes is increased. If the electrode is placed off center in a rotated assembly, a considerable enhancement of current can be observed at high rotation rates. This has been shown for both a band and a disk, and has been used to probe electron-transfer rates between polymer films and solution species [148].

E. Exploitation of the Small Size of Ultramicroelectrodes

An obvious feature of ultramicroelectrodes is that they are useful for sampling chemical concentrations in small volumes or at discrete, microscopic locations. The desire to make chemical measurements inside the living brain is a clear example of where the small size of these chemical probes is vital.   This subject has been reviewed extensively [149,150] and is beyond the scope of this chapter.

However, there are several other examples where the small size of these electrodes has been exploited.   Baranski has described the use of anodic stripping voltammetry for cadmium and lead in volumes as small as 5 μl [151,152] (Fig. 24).   In addition to the advantage of detecting small volumes, there are two other advantages in anodic stripping voltammetry at ultramicroelectrodes that occur as a direct result of their small size [15].   First, stirring is not required during the deposition step because of the enhanced steady-state diffusion of species to the electrode.   This minimizes one of the sources of error in the experiment.

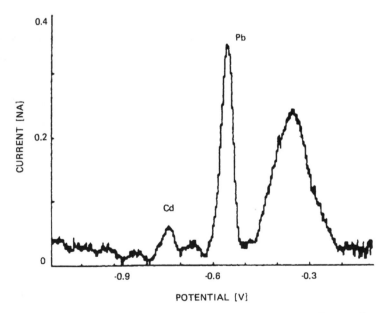

FIG. 24.   Subtractive stripping voltammogram of 5×10^{-8} M Pb^{2+} and 5×10^{-9} M Cd^{2+} in 5 μl of a nondeoxygenated 0.01 M HCl solution obtained with a Hg-coated carbon-fiber ultramicroelectrode.   Preconcentration sweep rate was 10 V/sec.   (From Ref. 152.)

Second, the analyte, which is accumulated in the very small volume of the liquid electrode, is completely stripped during the anodic scan. This leads to very sharp peaks, similar to those observed with mercury thin films. Ultramicroelectrodes have helped generate new insights into nucleation processes as well [12,153]. This is because the number of possible nucleation sites is reduced because of the small size of the electrode [153].

When an ultramicroelectrode is attached to a micromanipulator, the position of the electrode can be physically moved. In this way, heterogeneities in chemical concentrations can be examined. A carbon-fiber electrode encased in a small glass capillary has been used to probe heterogeneity at large electrode surfaces. By rastering the ultramicroelectrode through the diffusion layer of the larger electrode, a two-dimensional view of the electrochemical processes can be obtained with a resolution of 1×10^{-3} cm [154]. Carbon-fiber electrodes have been used to measure the radial dispersion which occurs during laminar flow inside tubes (r = 0.4 mm). With the use of a voltammetric electrode with a total diameter of 2×10^{-3} cm, the stagnant region near the wall of the tube could be clearly observed [155]. The small size of carbon-fiber cylindrical electrodes makes them ideal for use as chromatographic detectors with very small capillary columns [156]. Well-defined voltammograms can be obtained during elution of peaks from the column, and this aids in compound identification [157–159]. The ultimate in "small size" applications is to combine the principles of scanning tunneling microscopy with electrochemical measurements [160]. In this way, spatial and electrochemical measurements could be made with submicrometer resolution.

The smallest voltammetric electrodes that have been reported to date have the geometry of a band, and thus are microscopic in only one dimension. Bands with widths as small as 2 nm have been reported [161]. It is unlikely that thinner bands will be useful because metal films of

smaller dimensions consist of islands of metal, and do not conduct. The
precise dimensions of these electrodes are uncertain because microscopic
examination of these surfaces is difficult. Nevertheless, electrodes of
such small dimensions are of fundamental interest because they approach
molecular dimensions. In a voltammeric characterization of these electrodes, White reported that band electrodes with a width of less than
50 nm gave voltammetric limiting currents that were less than expected
from the equations given in Sec. II [161]. This result is unlikely to be
an artifact, since we have observed the same behavior with microbands
of similar dimensions [91]. White has proposed that this deviation from
diffusion-controlled theory is because as the diffusion layer dimensions
are decreased by reducing the electrode dimensions, the finite size of
the electrolyzed species decreases the concentration gradient at the
electrode surface. This occurs as a result of the restricted volume in
the diffusion layer. In addition, he notes that diffusion coefficients
are likely to decrease in the structured region of solvent adjacent to the
electrode. Although other factors may also play a role, the experimental
data suggest that care must be used when the data from nanometer-sized
electrodes are fit to the diffusion equations given earlier.

Closely spaced ultramicroelectrodes operated at different potentials
allow the study of the reaction of unstable products electrogenerated at
adjacent electrodes. For example, cation-anion annihilation reactions can
occur in the space between adjacent bands operated at opposing potentials,
with a concomitant generation of light [162]. If the electrogenerated
product undergoes a chemical reaction before reaching an adjacent band,
a lower collection efficiency will be obtained. Since the time for a
reagent to diffuse from one adjacent band is approximately equal to the
square of the distance between the bands [83], very fast chemical
reactions can be examined if the bands are placed sufficiently close.
This has been demonstrated for the catalytic reduction of oxygen by a
water-soluble hydroquinone, and for the catalytic oxidation of ascorbic

acid and aminopyrine by ferricyanide [163]. Digital simulation of the reaction sequence has been developed so that quantitative rate constants can be obtained.

The interaction between closely spaced ultramicroelectrodes is much like the events that occur in thin-layer cell with twin, opposing electrodes. If the adjacent electrodes are coated with a conducting polymer, the current is primarily affected by the mass transport processes in these films [162,164—166]. This provides a unique way to study these mass transport processes. With both the working and counter electrode in the film, the electrochemical cell is complete and a liquid phase is not required. It has been shown that a cell operated in this way is sensitive to the specific gas that surrounds the cell, presumably because the gases act as plasticizers and alter the film in different ways [167]. This type of device has significant promise as an analytical sensor.

Adjacent ultramicroelectrodes coated with polymer films can be used to perform functions similar to those of solid-state electronic devices [168,169]. An early example was a molecular transistor. The electrical communication between two adjacent bands was controlled by adjusting the redox state and thus conductivity of a film coated over both electrodes by a third, external electrode [170]. Although the devices constructed to date operate on a much slower time scale than solid-state devices, their chemical basis suggests that they, too, will be useful as chemical sensors. It has been shown that the frequency response can be improved by reducing the size of the gap between adjacent electrodes [171].

Thus it is clear that electrodes with microscopic dimensions have many unique advantages. The small size of these electrodes reduces limitations caused by the double-layer capacitance and solution resistance. Their small size also facilitates examination of chemical phenomena across very small dimensions. As these advantages become more clear, and the theory of processes at these electrodes becomes more widely understood, their use in chemical applications should continue to grow.

IV.   CONSTRUCTION AND USE OF ULTRAMICROELECTRODES

Several materials were developed in the 1960s and 1970s which provided
the materials for building ultramicroelectrodes.   For example, carbon
fibers, which were among the earliest used materials for ultramicroelectrodes [3,172,173], are available commercially with radii in the micrometer range.   These fibers are composed of almost pure carbon and are
highly conductive.   Microscopic wires and films of noble metals are also
commercially available and can be used to fabricate ultramicroelectrodes.
The development of microlithography techniques has also provided a
route to prepare ultramicroelectrodes.   The concurrent development in
electronic devices has facilitated the measurement of very small currents
which are often encountered with ultramicroelectrodes.   Thus developments in a number of different areas made feasible the construction and
use of ultramicroelectrodes.

   In this section, methods of fabrication of different types of ultramicroelectrodes are described.   The construction methods that are
described are primarily the ones currently used in our laboratory.
Several other designs are described in the literature, and it is anticipated that many more sophisticated designs will appear in the future.
Certainly, microlithography techniques will play an important role in
future ultramicroelectrode advances [83,84,90,102,162,170,174−176].
The designs described in the following pages were arrived at by considering several criteria.   The electrode should be capable of use in a
wide range of solvents, it should be simple to construct with dimensions
that can be accurately estimated, and it should be reusable.   The
criteria of reusability requires that methods of surface cleaning be
developed.   We have adopted conventional polishing techniques.   The
use of polishing mitigates against many forms of photolithography, since
the structures created in this way may be destroyed by polishing.

A.   Construction of Robust Disk-Shaped Ultramicroelectrodes

Disk electrodes can be prepared by sealing fine wires into glass tubes
(Fig. 25) [5].   Glass tubes serve as the insulator and body of the

electrode. Individual insulators are made from soft glass tubing (4 mm o.d. and 2.5 mm i.d), and are approximately 10 cm long. One end of the insulator is drawn to a taper with an o.d. of 1.5 mm and i.d. of 1—0.5 mm with a taper length of about 7 mm. It is important that the walls of the tapered end are not thinner than that of the original glass. To prepare 10-μm-diameter gold and platinum disk electrodes, wires of these dimensions are employed (obtained from Goodfellow Metals, Cambridge, UK). Since the wires are fragile and difficult to see, the

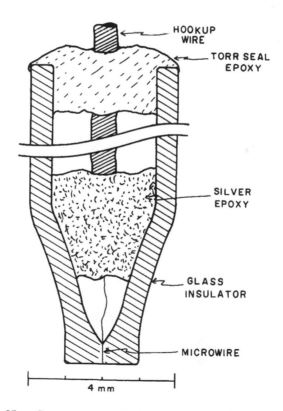

FIG. 25. Cross-sectional view of a microdisk electrode.

electrode is prepared on a clean white surface with a strong light source at hand. Approximately 1 cm of wire is inserted into the tapered end of the glass insulator with fine tweezers. Bends in the wire should be avoided because this makes insertion into the insulator difficult. Gold wire is more of a problem in this regard. The wire will remain in position in the insulator unless rough handling dislodges it.

The wire is sealed into the glass by melting the glass. The preferred method is to use a resistance heater with controllable current (15–20 A ac) with the use of coiled nichrome wire, because the heat can easily be controlled. The coil diameter should be approximately twice that of the tapered end of the electrode blank. The glass insulator containing the electrode wire is positioned so that the tapered end enters the center of the coil with a micromanipulator (occasionally, the coil shifts position during heating, so its position should be checked after the coil is hot). The glass should melt around the wire for 1–2 mm at the tip of the assembly. After the glass has cooled, the electrode should be inspected under a microscope to determine if the wire is completely sealed in the tip and if there are any trapped air bubbles.

The preparation of disk-shaped electrodes of 2 and 0.6 µm diameter is similar to that described above. The major difference is that the wire is supplied in the form of Wollaston wire (Goodfellow Metals, Cambridge, UK) [15,138], i.e., it has a 50- to 100-µm silver coating surrounding the noble metal. The first step is to cut a 1.5-cm length of Wollaston wire and shape it into a hook with a 1-cm shank. The wire is simpler to handle than the 10-µm wires because of the silver coating. The glass insulators employed are similar to those described above but have a narrower and slightly longer taper. The wire is placed in the large end of an insulator and forced down the shaft with a wooden stick until the shank of the hook is in the tapered region. The hook acts as a spring to hold the wire in place.

The silver coating is removed with nitric acid. The tip of the insulator containing the wire hook is dipped into a vial of 1:1 HNO_3.

The silver dissolves at a rate that is inversely proportional to the inside diameter of the taper. Because of the greater control involved with slower etching, a narrow bore is preferred. The silver is etched until 1−2 mm of the platinum wire is exposed. The nitric acid is removed with the aid of a paper tissue, and the electrode tip is dipped into distilled water. These steps are repeated until the acid is removed, and then the tip is rinsed with acetone and left to dry. Care must be taken during the etching procedure because the exposed platinum wire is very fragile and easily broken.

The wire is sealed into the glass insulator with the use of the same procedure as with the larger electrodes. However, it is important that the location where the platinum joins the silver is sealed in glass; otherwise, the platinum wire will break during the polishing steps. To accomplish this, the junction must be within 1−2 mm of the tip of the glass. The wire can be manipulated so that this condition is met by filling the barrel of the insulator with acetone and pushing the hooked end farther into the barrel with a wooden stick.

The method for making sealed-in-glass carbon fiber electrodes is the same as that for the construction of 10-μm-diameter gold or platinum electrodes. We generally employ 10-μm-diameter carbon fibers prepared from the pyrolysis of pitch (Thornell VSB-32, type P, Union Carbide Corp., New York, NY). The sealing procedure must be done in an inert atmosphere such as a glove bag, dry box, or under vacuum because CO_2 is evolved and a poor seal results.

When a satisfactory seal is obtained the disk is exposed by polishing. Polishing also removes bubbles present in the glass or other defects. When a large amount of insulator must be removed, 600-grit carborundum should be used. A small amount of the dry material is placed in a glass plate and water is added to form a suspension. The electrode is polished by grinding with the electrode held perpendicular to the surface, and a figure eight motion is followed with periodic rotation of the barrel of the

electrode.   Alternatively, the plate can be attached to a polishing wheel
and polished by holding the electrode perpendicular to the wheel.   The
carborundum removes material very rapidly, so the progress should be
frequently checked.

Following the coarse step, the electrode is polished with 9-μm paper
using the same figure eight motion as used above until the surface is
flat and shows uniform scratches under the microscope.   Next, the
surface is polished on a polishing wheel, with the use of a polishing
cloth and 5-μm alumina (Beuhler, VWR Scientific), until the surface looks
mirror smooth to the eye (this step may take 20 min).   Under the
microscope the surface will appear to be uniformly scratched.   If large
gouges are present, additional polishing must be done.   Next, the
surface is polished with 1-μm alumina for 10 min to make the scratches
less deep.   (It is essential to use a different cloth for every grade of
alumina polish used.   The back of the cloth should be marked with an
indelible pen to indicate the grade of polish used.)   The polishing is
continued with 0.3-μm alumina for 10 min.   After this step, scratches
will be apparent only at high magnification.   The last step is to polish
with 0.05-μm alumina for 10 min.   Then the surface will be mirror smooth
and without scratches.

After the wire is sealed and the electrode is polished, electrical
contact is made.   "Hookup" wire [22-gauge Teflon coated wire (Belden)]
is cut to 6-in. lengths, the insulation is removed from each end, and
the ends are tinned with rosin-core solder.   We use different-colored
wires for connection to different electrode materials: yellow for Au
electrodes, white for Pt electrodes, black for carbon-fiber electrodes.
A silver epoxy (Epo-Tek H20E, Epoxy Technology, Inc., Billerica, MA)
is employed to hold the hookup wire in place.   The conductive epoxy is
inserted into the insulator with a disposable syringe so that the epoxy
is in intimate contact with the electrode wire and the contact wire is
inserted down the barrel into the epoxy.   The epoxy is cured at 100°C

for at least 2 hr. After this, a small amount of Torr Seal epoxy (Varian Associates, Vacuum Division, Palo Alto, CA) is packed into the open end so that the contact wire is completely embedded. This seals the barrel and provides strain relief for the contact wire.

B. Beveled Carbon-Fiber Electrodes

Electrodes with a small outer insulation are used in applications where a small chemical sensor is required. These electrodes are fragile and are usually discarded after 1 day's use. This style of electrode is prepared by aspirating carbon fibers of 5 μm radius into glass capillaries of 1.0 mm o.d. containing an inner filament [177]. The capillaries are then pulled in a standard electrode puller (Model PE-2, Narishige Scientific Instruments, Tokyo) and the glass is tapered around the fiber. The tip is then cut with a scalpel so that the glass at the tip has an approximate diameter of 15 μm. The fiber is sealed in the capillary using epoxy (Epon 828, Shell Chemical Company, Houston, TX) with 14% by weight metaphenylene diamine as hardening agent (Miller-Stephenson Chemical Co., Danbury, CT; they will also sell the Epon 828) and cured at 150°C overnight. The epoxy will fill the electrode tip by capillary action. This step is necessary if the capillaries are pulled in air because a glass-carbon seal will not be achieved.

To meet the criteria of chemical stability, the selection of epoxy for the sealing step is critical. Most "hardware store" epoxies are very susceptible to chemical attack and are therefore unacceptable. However, the epoxy mixture described above is an almost ideal material. The amine does not dissolve in the epoxy at room temperature, but must be heated to 70°C. The mixture hardens at 80°C, so care must be taken in the heating step. At 70°C the epoxy is virtually inviscid and will flow into very small cracks. The epoxy should be cured for at least 2 hr at 150°C. The cured material is remarkable because it is insoluble in all organic solvents that we have tried. For example, voltammograms can be recorded for several hours in acetonitrile wth no sign of failure. Even dimethyl-formamide does not seem to attack the epoxy. It is soluble in 5 N nitric acid.

The electrodes are mounted on a tilt-base micromanipulator for the beveling step (Model 55133, Stoelting Co., Chicago) at an angle of approximately 45°. Because of the fragility of these electrodes, they cannot be polished by hand. A thin layer of diamond polishing compound (Metadi II, 1-μm particle size, Buehler Ltd., Lake Bluff, IL) is spread over the reflective glass surface of a microelectrode beveler (Model 1300, World Precision Instruments, New Haven, CT) which has been modified to rotate at a rate of 180 rpm. This glass surface is optically flat to within 2.5 μm and is coated with 0.05-μm alumina powder embedded in urethane. The tip of the electrode is lowered with the use of the coarse control of the micromanipulator until it is just above the surface of the diamond polish. The wheel is then turned on, and using the manipulator fine control (10 μm calibration), the microelectrode is lowered until the tip just touches the surface polish. After making a reading of the micromanipulator position, the electrode is lowered between 50 and 100 μm farther and left in contact with the rotating wheel for a period of 10 min. This final distance that the electrode is moved depends on the roughness of the tip before polishing and the thickness of the layer of diamond polish. After polishing, the electrode is carefully retracted from the glass plate (with the wheel still turning) and dipped in toluene at a temperature just below the boiling point for 15−30 sec to remove any polishing compound that remains on the electrode tip.

Once polished, the tip of the microelectrode should be smooth and free from cracks. The length of the bevel can be changed as desired by controlling the angle of incidence to the polishing wheel or the amount of pressure with which the electrode is held against the wheel during polishing.

C. Construction of Mercury Hemispherical Electrodes

The construction of mercury ultramicroelectrodes involves growing a hemispherical deposit of mercury onto a platinum substrate [15]. A previously constructed platinum-disk microelectrode (0.6, 2.0, or 10 μm in diameter) is polished with 0.05-μm alumina and placed in a deoxygenated solution of Hg(I) ions with an applied potential of 0.0 V vs. SCE. The selection of the substrate is critical because many materials will amalgamate with mercury. Platinum can amalgamate [178], but oxide films retard this process. Iridium is insoluble in mercury and is thus a far better choice [179]. Another alternative is carbon [12]. However, it is difficult to grow stable and reproducible films on carbon [180].

The deposition solution contains 1.0 M KNO_3 with 5.7 mM mercurous ion and 0.5% HNO_3. The current is monitored with a strip-chart recorder. The radius of the hemispherical deposit (r) is related to the deposition time by

$$r = \left(\frac{2MDCt}{d} \right)^{1/2} \tag{52}$$

where M is the atomic weight of mercury (200.59 g/mole, D the diffusion coefficient of the mercurous ion (9.6×10^{-6} cm^2 sec^{-1}), C the concentration of the mercurous ion, t the deposition time in seconds, and d the density of mercury (13.55 g cm^{-3} at 25°C). The mercury is deposited so that the radius of the drop is larger than the <u>diameter</u> of the platinim substrate since the exact location of nucleation is not known. The exact radius can be verified by testing as described in a subsequent section. After use, the mercury drop can be removed by polishing with 0.05-μm alumina.

D. Construction of Cylindrical Electrodes

Platinum microcylinder electrodes can be constructed by sealing 5-μm-radius platinum wire (Goodfellow Metals, Ltd., Cambridge, England) in glass borosilicate capillary tubes with an electric heating coil [57].

Electrical contact to the platinum wire can be made with mercury. The
platinum wire is usually trimmed so that the cylinder is approximately
0.5 mm long. The cylindrical surface cannot be polished mechanically,
so electrochemical pretreatments must be used.

Microcylindrical electrodes can also be constructed with the use of
carbon fibers [3,67]. Individual fibers are inserted into glass capillaries
by aspiration. The glass tube is then pulled to a fine taper around the
fiber with a pipette puller as used with the beveled carbon-fiber electrodes. A tight seal is achieved by carefully placing a drop of epoxy
(Epon 828) at the end of the glass. This is accomplished by placing
another pulled capillary, filled with epoxy, in close proximity to the
electrode. The epoxy is applied by positive pressure on the open end
of the capillary, and care is taken so that the epoxy enters the electrode capillary without coating the carbon fiber that extends beyond the
glass. The fiber is trimmed with a scalpel so the approximately 0.5 mm
extends beyond the end of the capillary. The exact length of the pro-
truding fiber should be measured with an optical microscope. Electrical
contact to the wire or fiber can be made with mercury or silver epoxy.

Voltammetry at carbon-fiber, cylindrical electrodes tends to be
characterized by sluggish electron transfer rates, which is similar to
that seen at other forms of unactivated carbon [94,181]. For example,
the peak potential for ascorbate oxidation tends to be approximately 400
mV more positive at carbon than at mercury electrodes. However,
electrochemical treatment of the electrode results in a surprising increase
in the apparent kinetics, with accompanying shifts in the peaks for the
oxidation of a number of substances. The activation procedure most
widely used is a 0- to 3.0-V triangular wave at a frequency of 70 Hz,
and was introduced by Gonon and co-workers [94,182]. Ths pretreatment is widely used by investigators of in vivo electrochemistry because
it gives well-shaped separated peaks for ascorbate and dihydroxyphenlyl-
acetic acid, two substances of neurochemical importance. Furthermore,

the activated electrode retains its characteristics when implanted in the hostile environment of the brain for several hours. The activation procedure is much more difficult to reproduce at electrodes of conventional size because a current density of over 3 A cm^{-2} is required.

Although the exact mechanism that results in the activation is not known, the properties of this electrode have been widely explored. Electrochemical pretreatment increases the oxides of carbon on the surface, and these can serve as anchors for surface modification [183,184]. It has been shown that cations such as cupric ion [185] and catecholamines [67,96,186], which possess a protonated amine on their aliphatic side chain, and both ruthenium and cobalt hexaammine adsorb to the electrode [67]. Although adsorption was difficult to evaluate in the early studies of these activated electrodes, the developments in the theory of amperometry and voltammetry at cylindrical electrodes clearly showed that the measured current was larger than expected for diffusion control. In contrast, anions such as ascorbate, ferricyanide, and molybdenum(IV) octacyanide all give current that is less than expected for a cylindrical electrode with the geometric area of the electrochemically modified cylinder [67]. This has been interpreted to mean that a large portion of the surface is blocked by an insulating film, and the electrochemistry occurs only at small regions on the electrode. Surface analysis of the carbon fibers after electrochemical treatment shows the presence of large amounts of surface oxides and the presence of surface cracks [187,188], which supports the electrochemical interpretation of the data. Nevertheless, the chemical factors that cause the apparent increase in rate in activated surfaces is still unclear.

Mercury can be coated onto carbon-fiber cylindrical electrodes by electrodeposition techniques [189–191]. However, the mercury films tend not to be uniform. This type of electrode is useful for anodic [189,192] or potentiometric [146,193–196] stripping analysis.

E. Construction of Band Ultramicroelectrodes

Single-band electrodes may be constructed by sealing thin films of gold
or platinum in an insulating material [57,91]. Platinum foil of 4 and 10
μm thickness and gold foil are available from Aesar, Johnson Matthey
Inc., Seabrook, New Hampshire. A variety of thinner foils are available
from Goodfellow Metals, Cambridge, UK. The thin foils come on a
variety of support materials. Mylar is preferred because it is chemically
inert. Alternatively, thin films can be prepared locally by standard
procedures for preparing thin metal or carbon films on insulating substrates such as mica [161,163]. We have used two methods to seal the
foils: melting soft glass around the films or the use of epoxy to attach
the insulator. The best seal is obtained with glass, but this method
can only be used with thick foils (approximately 4 μm and greater).
The epoxy procedure is amenable to all types of single-band electrodes
and paired-band electrodes.

To seal the foils into glass, soft glass tubes, 4 mm i.d. and about
10 cm long, are employed. The end of the glass tubing is softened in
a bunsen burner flame. A pair of flat brass tongs is used to form the
glass tip into a slot that will accommodate a piece of the electrode
material. A strip of the foil, about 5 by 10 mm, is cut with a scalpel.
The short edge will form the electrode surface. The strip is handled
with tweezers and slid into the slot of the glass tube. The end is
reheated at a higher temperature to obtain the seal. The tongs are
used to close the end of the tube completely around the metal foil. The
edges are generally more difficult to seal, so extra attention must be
paid there.

When the foil is sealed into the glass, the end is sanded flat with a
mechanical sander to reveal the electrode edge. A connecting wire is
coated with silver epoxy and inserted into the open end of the tube to
make contact with the foil and then cured at 100°C for 30 min. Torr
Seal is used to seal the open end and to provide support for the contact wire.

The band electrodes prepared with epoxy employ glass-microscope slides as the insulating material (Fig. 26). Two pieces of glass are used, one 1 cm × 2 cm and the other 1 cm × 1 cm, with a strip of foil about 0.5 cm × 1.5 cm. The electrode is constructed on a Teflon sheet because epoxy will not stick to this surface. The larger glass piece is taped down at the edges, and the metal foil is placed on the glass. A small amount of the Epon 828 epoxy, prepared as described earlier, is applied to both sides of the foil. The smaller piece of glass is placed on top of the foil, sandwiching the foil between the glass pieces. Enough epoxy is needed to make the seal between the glass pieces and foil complete, but too much will cover the bottom of the foil where elec-

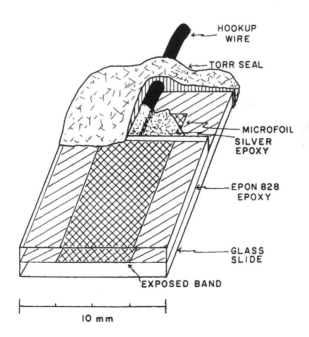

FIG. 26. Sandwich-type band ultramicroelectrode with a cutaway view showing the electrical connection.

trical contact will be made. The assembly is clamped together with a large spring clip and the epoxy cured at 175°C for 2 hr. After curing, a lead wire is attached to the exposed portion of the metal film with the use of silver epoxy. The exposed wire, silver epoxy, and electrode sides are covered with Torr Seal to prevent solution contact. After the Torr Seal hardens, the metal band is exposed.

Electrodes that contain two parallel bands spaced by a few micrometers of insulating material are useful for the collector-generator experiment. Construction of these electrodes can be accomplished with epoxy in a similar way to that for single bands. However, a Mylar film of thickness 2 to 4 μm (Spectrofilm, Polaron Instruments, Hatfield, PA) is placed in between the two pieces of the foil in the sandwich assembly. A piece of Teflon film is used to prevent a short between the two foils in the region outside the sandwich. After curing the bands should still be aligned, and there should be access to both strips of foil. Electrical contact is made to each electrode as before, the excess Mylar and Teflon films are trimmed, and the sides are covered with Torr Seal. The electrode is then sanded until the paired bands are revealed.

The band electrodes are first polished on a glass plate with a suspension of 1000-grit carborundum and water. Next, the electrode is polished on a cloth with cerium oxide (1 μm particle size, Buehler) which has been diluted three times with water. A polishing wheel is employed, and the electrode should be polished for 10–20 min. Careful polishing with the cerium oxide results in a surface free of scratches when viewed under a microscope at a high magnification. Alumina has been found not to be useful as a polishing material with band electrodes. The band tends to become recessed inside the insulator as evidenced by a low limiting current and voltammograms that have the shape expected for linear diffusion. If further polishing is desired, 0.25-μm diamond paste on a clean cloth should be used.

F. Miscellaneous Ultramicroelectrodes and
Arrays of Ultramicroelectrodes

Several other types of ultramicroelectrodes have been described. Electrodes with a ring geometry have been constructed by MacFarlane and
Wong [197]. These were prepared by vapor deposition or sputtering of
metal onto a glass rod. The rod is rotated during deposition to ensure
a thin film. Films ranging from 30 to 400 nm have been used [48,91,197].
The film is then insulated from solution by sealing it into a glass tube
of slightly larger dimensions. The ring geometry is then exposed by
cutting the sealed assembly in a direction perpendicular to the axis of
the rod. Carbon rings can be formed in a similar way by the pyrolysis
of a methane jet through a small capillary. Carbon forms on the inner
walls, and the center is filled with epoxy [198].

A disk electrode sufficiently small for intracellular measurements has
been prepared [199]. A carbon fiber was etched electrochemically in
sodium nitrite containing Tween 40 with the use of alternating current.
Then the tip of the electrode was sealed in glass to give a tip of 1 μm.
Conical electrodes have been prepared by electrochemically etching
carbon fibers in dichromate solutions [200].

Well-defined arrays of ultramicroelectrodes are probably best constructed with microlithography techniques. Alternatively, they can be
constructed by combining a large number of metal wires or carbon fibers
[72,146,180]. For analytical purposes, a defined array is often unneces-
sary because a theoretical description of the current is not required
since calibration curves will be employed. This was illustrated with the
use of reticulated vitreous carbon filled with epoxy [201]. This carbon
material is porous, and when filled with epoxy, the surface contains a
large number of exposed islands of carbon. This design exhibits many
of the voltammetric features expected for an array of ultramicroelectrodes.

To accomplish large-scale electrolysis with ultramicroelectrodes, dispersions of small platinum particles within a large electric field gradient

have been used [202,203]. The particles are kept in suspension by
convection. The potential gradient across each particle appears to
enable it to act as both an anode and a cathode.

G. Instrumentation

Since the current from ultramicrodisks, spheres, and rings is small,
commercial electrochemical instrumentation designed for use with conventional electrodes is not suitable for most measurements. Therefore,
most investigators have constructed their own devices to measure the
low currents. For low-frequency applications, a current follower employing an AD515K operational amplifier (Analog Devices) is ideal [93]. To
adapt conventional instrumentation for use with ultramicroelectrodes, a
simple, two-amplifier system has been developed [204]. The device
consists of a hgh-gain current follower coupled with an inverter so that
the sign of the output voltage will be compatible with the commercial
instrument.

Because the effects of solution resistance are much less severe at
ultramicroelectrodes, three-electrode potentiostats are not required.
Two-electrode systems suffice as long as the reference electrode is of
low impedance [132]. Thus for low-frequency applications the potential
waveform can be applied directly to the reference electrode, and a
commercial picoammeter can be used to measure the current [205,206].
For very fast measurements, locally constructed instruments are required [136].

With band and cylinder ultramicroelectrodes, as well as with arrays,
the currents tend to approach the microampere level for typical operating
conditions. In this case commercial instrumentation can be employed.

H. Testing Procedures

Evaluation of newly constructed ultramicroelectrodes can be accomplished
by comparing their limiting current to the appropriate expressions under
steady-state conditions ($D_0 t / r_0^2 > 100$). For most of the sizes of

electrodes described here, cyclic voltammetry at a scan rate of 100 mV sec^{-1} will suffice. Alternatively, constant potential may be used if a flow injection apparatus is available. Electrodes are tested with either the reduction of ruthenium hexaammine (1.0 mM $Ru(NH_3)_6Cl_3$ in a pH 7.4 phosphate buffer consisting of 60 mM Na_2HPO_4 and 30 mM $Na_2HPO_4 \cdot H_2O$) or the reduction of ferricyanide [1.0 mM $K_3Fe(CN)_6$ in either the phosphate buffer above or in 0.5 M K_2SO_4 adjusted to pH 3.0 with H_2SO_4]. Platinum, gold, and carbon can all be tested with success in these media. Mercury electrodes work well with ruthenium hexaammine. The electrodes should give a sigmoidal response in accordance to theory. The baseline should be flat with no significant slope except for the charging current. The magnitude of the limiting current to that given by theory should agree to within 3—5%. The diffusion coefficient for the ruthenium hexaammine is 5.4×10^{-6} cm^2 sec^{-1} in pH 3.0 sulfate, and 5.3×10^{-6} cm^2 sec^{-1} in the phosphate buffer. These diffusion coefficients were calculated using the expression for steady-state current at a disk [Eq. (18)]. Electrode radii were verified using electron microscopy.

If no response is seen, the electrode should be repolished and tried again. If there is still no response, the electrode should be discarded. If the background current during cyclic voltammetry is ramped or a large charging current is observed, the seal of the electrode is poor. Various other transient electrochemical techniques can be used to characterize further poorly sealed electrodes [207]. A leaky electrode is caused by bubbles at the tip or a gap between the electrode and the sealing material and can often be observed under the microscope. In some cases the electrode can be made serviceable by repeating the polishing steps to remove a small amount of the tip. The beveled carbon-fiber electrodes have an elliptical surface, and exact theory does not exist for that geometry. However, we find experimentally that the response can be predicted by considering a disk of equivalent area to the ellipse [177].

ACKNOWLEDGMENTS

Research in our laboratory in the area of ultramicroelectrodes has been supported by the National Science Foundation. The comments of J. L. Anderson, J. O. Howell, and A. Szabo are gratefully acknowledged.

REFERENCES

1. P. N. Swan, Ph.D. Dissertation, University of Southhampton, 1980.

2. R. N. Adams, Anal. Chem. *48*:1126A (1976).

3. J. -L. Ponchon, R. Cespuglio, F. Gonon, M. Jouvet, and J. -F. Pujol, Anal. Chem. *51*:1483 (1979).

4. R. M. Wightman, Anal. Chem. *53*:1125A (1981).

5. J. O. Howell and R. M. Wightman, Anal. Chem. *56*:524 (1984).

6. C. A. Amatore, A. Jutand, and F. Pfluger, J. Electroanal. Chem. *218*: 361 (1987).

7. J. Ghoroghchian, F. Sarfarazi, T. Dibble, J. Cassidy, J. J. Smith, A. Russell, G. Dunmore, M. Fleischmann, and S. Pons, Anal. Chem. *58*:2278 (1986).

8. M. A. Dayton, J. C. Brown, K. J. Stutts, and R. M. Wightman, Anal. Chem. *52*:946 (1980).

9. A. J. Bard and L. R. Faulkner, *Electrochemical Methods*, Wiley, New York, 1980.

10. D. MacGillavry and E. K. Rideal, Recl. Trav. Chim. Pays-Bas *56*:1013 (1937).

11. P. Delahay, *New Instrumental Methods in Electrochemistry*, Inter-Science, New York, 1954.

12. B. Scharifker and G. Hills, J. Electroanal. Chem. *130*:81 (1981).

13. I. Shain and K. J. Martin, J. Phys. Chem. *65*:254 (1961).

14. M. Ikeuchi, Y. Fujita, K. Iwai, and G. P. Sato, Bull. Chem. Soc. Jpn. *49*:1883 (1976).

15. K. R. Wehmeyer and R. M. Wightman, Anal. Chem. *57*:1989 (1985).

16. R. S. Nicholson and I. Shain, Anal. Chem. *36*:706 (1964).

17. Z. Galus, *Fundamentals of Electrochemical Analysis*, Ellis Horwood, Chichester, West Sussex, England, 1976.

18. K. B. Oldham, J. Electroanal. Chem. *122*:1 (1981).

19. Z. Galus, J. O. Schenk, and R. N. Adams, J. Electroanal. Chem. *135*:1 (1982).

20. J. Albery and S. Bruckenstein, J. Electroanal. Chem. *144:*105 (1983).

21. J. Newman, J. Electrochem. Soc. *113:*501 (1966).

22. Z. G. Soos and P. J. Lingane, J. Phys. Chem. *68:*3821 (1964).

23. J. B. Flanagan and L. Marcoux, J. Phys. Chem. *77:*1051 (1973).

24. A. J. Bard, Anal. Chem. *33:*11 (1961).

25. P. J. Lingane, Anal. Chem. *36:*1723 (1964).

26. C. R. Ito, S. Asokura, and K. Nobe, J. Electrochem. Soc. *119:*698 (1972).

27. Y. Saito, Rev. Polarogr. Jpn. *15:*177 (1968).

28. H. S. Carslaw and J. C. Jaeger, *Conduction of Heat in Solids* (2nd Ed.), Clarendon Press, Oxford, 1959.

29. D. Shoup and A. Szabo, J. Electroanal. Chem. *160:*27 (1984).

30. K. Aoki and J. Osteryoung, J. Electroanal. Chem. *122:*19 (1981).

31. D. Shoup and A. Szabo, J. Electroanal. Chem. *140:*237 (1982).

32. K. Aoki and J. Osteryoung, J. Electroanal. Chem. *160:*335 (1984).

33. M. Kakihana, H. Ikeuchi, G. P. Sato, and K. Tokuda, J. Electroanal. Chem. *108:*381 (1980).

34. M. Kakihana, M. Ikeuchi, G. P. Sato, and K. Tokuda, J. Electroanal. Chem. *117:*201 (1981).

35. T. Hepel and J. Osteryoung, J. Phys. Chem. *86:*1406 (1982).

36. T. Hepel, W. Plot, and J. Osteryoung, J. Phys. Chem. *87:*1278 (1983).

37. K. Aoki and J. Osteryoung, J. Electroanal. Chem. *125:*315 (1981).

38. A. S. Baranski, W. R. Fawcett, and C. M. Gilbert, Anal. Chem. *57:*166 (1985).

39. A. M. Bond and K. B. Oldham, J. Electroanal. Chem. *158:*193 (1983).

40. J. Heinze, Ber. Bunsenges. Phys. Chem. *85:*1096 (1981).

41. B. Speiser and S. Pons, Can. J. Chem. *60:*1352 (1982).

42. B. Speiser and S. Pons, Can. J. Chem. *60:*2463 (1982).

43. B. Speiser and S. Pons, Can. J. Chem. *61:*156 (1983).

44. J. F. Cassidy, S. Pons, A. S. Hinman, and B. Speiser, Can. J. Chem. *62:*716 (1984).

45. D. Shoup and A. Szabo, J. Electroanal. Chem. *160:*1 (1984).

46. K. Aoki, K. Akimoto, K. Tokuda, H. Matsuda, and J. Osteryoung, J. Electroanal. Chem. *171:*219 (1984).

47. K. Aoki, K. Akimoto, K. Tokuda, M. Matsuda, and J. Osteryoung, J. Electroanal. Chem. *182:*281 (1985).

48. M. Fleischmann, S. Bandyopadhyay, and S. Pons, J. Phys. Chem. *89:*5537 (1985).

49. M. Fleischmann and S. Pons, J. Electroanal. Chem. *222:*107 (1987).

50. A. Szabo, J. Phys. Chem. *91:*3108 (1987).

51. H. A. Laitinen and I. M. Kolthoff, J. Phys. Chem. *45:*1061 (1941).

52. T. Berzins and P. Delahay, J. Am. Chem. Soc. *73:*555 (1951).

53. M. M. Nicholson, J. Am. Chem. Soc. *76:*2539 (1954).

54. G. L. Booman, E. Morgan, and A. L. Crittenden, J. Am. Chem. Soc. *78:*5533 (1956).

55. O. R. Brown, J. Electroanal. Chem. *34:*419 (1972).

56. M. M. Stephens and E. D. Moorhead, J. Electroanal. Chem. *164:*17 (1984).

57. P. M. Kovach, W. L. Caudill, D. G. Peters, and R. M. Wightman, J. Electroanal. Chem. *185:*285 (1985).

58. K. Aoki, K. Honda, K. Tokuda, and H. Matsuda, J. Electroanal. Chem. *182:*267 (1985).

59. K. Aoki, K. Honda, K. Tokuda, and H. Matsuda, J. Electroanal. Chem. *186:*79 (1985).

60. K. Aoki, K. Honda, K. Tokuda, and H. Matsuda, J. Electroanal. Chem. *195:*51 (1985).

61. S. Sujaritvanichpong, K. Aoki, K. Tokuda, and H. Matsuda, J. Electroanal. Chem. *199:*271 (1986).

62. K. Aoki, K. Tokuda, and H. Matsuda, J. Electroanal. Chem. *206:*47 (1986).

63. A. Szabo, D. K. Cope, D. E. Tallman, P. M. Kovach, and R. M. Wightman, J. Electroanal. Chem. *217:*417 (1987).

64. R. S. Robinson and R. L. McCreery, Anal. Chem. *53:*997 (1981).

65. R. S. Robinson, C. W. McCurdy, and R. L. McCreery, Anal. Chem. *54:*2356 (1982).

66. C. A. Amatore, M. R. Deakin, and R. M. Wightman, J. Electroanal. Chem. *206:*23 (1986).

67. P. M. Kovach, M. R. Deakin, and R. M. Wightman, J. Phys. Chem. *90:*4612 (1986).

68. S. Coen, D. K. Cope, and D. E. Tallman, J. Electroanal. Chem. *215:*29 (1986).

69. K. Aoki, K. Tokuda, and H. Matsuda, J. Electroanal. Chem. *225:* 19 (1987).

70. C. Amatore, M. R. Deakin, R. M. Wightman, and B. Fosset, J. Electroanal. Chem. *225:*33 (1987).

71. M. R. Deakin, R. M. Wightman, and C. A. Amatore, J. Electroanal. Chem. *215:*49 (1986).

72. W. L. Caudill, J. O. Howell, and R. M. Wightman, Anal. Chem. *54:*2532 (1982).

73. T. Gueshi, K. Tokuda, and H. Matsuda, J. Electroanal. Chem. *89:* 247 (1978).

74. T. Gueshi, K. Tokuda, and H. Matsuda, J. Electroanal. Chem. *101:*29 (1979).

75. K. Tokuda, T. Gueshi, and H. Matsuda, J. Electroanal. Chem. *102:*41 (1979).

76. O. Contamin and E. Levart, J. Electroanal. Chem. *136:*259 (1982).

77. C. Amatore, J. M. Saveant, and D. Tessier, J. Electroanal. Chem. *147:*39 (1983).

78. C. Amatore, J. M. Saveant, and D. Tessier, J. Electroanal. Chem. *146:*37 (1983).

79. J. Cassidy, J. Ghoroghchian, F. Sarfarazi, J. J. Smith, and S. Pons, Electrochim. Acta *31:*629 (1986).

80. D. Shoup and A. Szabo, J. Electroanal. Chem. *160:*19 (1984).

81. H. Reller, E. Kirowa-Eisner, and E. Gileadi, J. Electroanal. Chem. *138:*65 (1982).

82. H. Reller, E. Kirowa-Eisner, and E. Gileadi, J. Electroanal. Chem. *161:*247 (1984).

83. A. J. Bard, J. A. Crayston, G. P. Kittlesten, T. Varco Shea, and M. S. Wrighton, Anal. Chem. *58:*2321 (1986).

84. W. Thormann, P. van den Bosch, and A. M. Bond, Anal. Chem. *57:*2764 (1985).

85. V. Y. Filinovsky, Electrochim. Acta *25:*309 (1980).

86. S. Moldoveanu and J. L. Anderson, J. Electroanal. Chem. *185:*239 (1985).

87. J. L. Anderson, T. Y. Ou, and S. Moldoveanu, J. Electroanal. Chem. *196:*213 (1985).

88. D. K. Cope and D. E. Tallman, J. Electroanal. Chem. *205:*101 (1986).

89. L. E. Fosdick and J. L. Anderson, Anal. Chem. *58:*2481 (1986).

90. L. E. Fosdick, J. L. Anderson, T. A. Baginski, and R. C. Jaeger, Anal. Chem. *58:*2750 (1986).

91. K. R. Wehmeyer, M. R. Deakin, and R. M. Wightman, Anal. Chem. *57:*1913 (1985).

92. W. F. Berry and S. G. Weber, J. Electroanal. Chem. *208:*77 (1986).

93. A. G. Ewing, M. A. Dayton, and R. M. Wightman, Anal. Chem. *53:*1842 (1981).

94. F. G. Gonon, C. M. Fombarlet, M. J. Buda, and J. F. Pujol, Anal. Chem. *53:*1386 (1981).

95. F. G. Gonon, F. Navarre, and M. J. Buda, Anal. Chem. *56:*573 (1984).

96. S. Sujaritvanichpong, K. Aoki, K. Tokuda, and H. Matsuda, J. Electroanal. Chem. *198:*195 (1986).

97. D. P. Whelan, J. J. O'Dea, J. Osteryoung, and K. Aoki, J. Electroanal. Chem. *202:*23 (1986).

98. J. O'Dea, M. Wojciechowski, J. Osteryoung, and K. Aoki, Anal. Chem. *57:*954 (1985).

99. S. A. Schuette and R. L. McCreery, J. Electroanal. Chem. *191:*329 (1985).

100. S. A. Schuette and R. L. McCreery, Anal. Chem. *58:*1778 (1986).

101. A. S. Baranski, J. Electrochem. Soc. *133:*93 (1986).

102. A. M. Bond, T. L. E. Henderson, and W. Thormann, J. Phys. Chem. *90:*2911 (1986).

103. D. Britz, J. Electroanal. Chem. *88:*309 (1978).

104. C. Kasper, Trans. Electrochem. Soc. *77:*353 (1940).

105. L. Nemec, J. Electroanal. Chem. *8:*166 (1964).

106. J. Newman, J. Electrochem. Soc. *117:*198 (1970).

107. C. Kasper, Trans. Electrochem. Soc. *77:*365 (1940).

108. S. Bruckenstein, Anal. Chem. *59:*2098 (1987).

109. A. M. Bond, M. Fleischmann, and J. Robinson, J. Electroanal. Chem. *168:*299 (1984).

110. A. M. Bond, M. Fleischmann, and J. Robinson, J. Electroanal. Chem. *172:*11 (1984).

111. M. R. Deakin, R. M. Wightman, and C. A. Amatore, J. Electroanal. Chem. *220:*49 (1987).

112. R. Lines and V. D. Parker, Acta Chem. Scand. *B31:*369 (1977).

113. J. O. Howell and R. M. Wightman, J. Phys. Chem. *88:*3915 (1984).

114. L. Geng, A. G. Ewing, J. C. Jernigan, and R. W. Murray, Anal. Chem. *58:*852 (1986).

115. L. Geng and R. W. Murray, Inorg. Chem. *25:*3115 (1986).

116. A. M. Bond and T. F. Mann, Electrochim. Acta *32:*863 (1987).

117. M. J. Pena and M. Fleischmann, J. Electroanal. Chem. *220:*31 (1987).

118. M. E. Philips, M. R. Deakin, M. V. Novotny, and R. M. Wightman, J. Phys. Chem. *91:*3934 (1987).

119. A. M. Bond, M. Fleischmann, and J. Robinson, J. Electroanal. Chem. *180:*257 (1984).

120. R. A. Malmsten, C. P. Smith, and H. S. White, J. Electroanal. Chem. *215:*223 (1986).

121. J. Cassidy, S. B. Khoo, S. Pons, and M. Fleischmann, J. Phys. Chem. *89:*8933 (1985).

122. T. Dibble, S. Bandyopadhyay, J. Ghoroghchian, J. J. Smith, F. Sarfarazi, M. Fleischmann, and S. Pons, J. Phys. Chem. *90:*5275 (1986).

123. J.-W. Chen and J. Georges, J. Electroanal. Chem. *210:*205 (1986).

124. M. Ciszkowska and Z. Stojek, J. Electroanal. Chem. *213:*189 (1986).

125. J. E. Davis and N. Winograd, Anal. Chem. *44:*2152 (1972).

126. T. E. Cummings, M. A. Jensen, and P. J. Elving, Electrochim. Acta *23:*1173 (1978).

127. D. Garreau and J. M. Saveant, J. Electroanal. Chem. *50:*1 (1974).

128. J. E. Mumby and S. P. Perone, Chem. Instrum. *3:*191 (1971).

129. R. S. Robinson and R. L. McCreery, J. Electroanal. Chem. *182:* 61 (1985).

130. J. O. Howell, J. Goncalves, C. Amatore, L. Klasinc, J. Kochi, and R. M. Wightman, J. Am. Chem. Soc. *106:*3968 (1984).

131. D. O. Wipf, K. R. Wehmeyer, and R. M. Wightman, J. Org. Chem. *51:*4760 (1986).

132. A. Fitch and D. H. Evans, J. Electroanal. Chem. *202:*83 (1986).

133. M. I. Montenegro and D. Pletcher, J. Electroanal. Chem. *200:*371 (1986).

134. K. R. Wehmeyer and R. M. Wightman, J. Electroanal. Chem. *196:* 417 (1985).

135. J. Millar, J. A. Stamford, Z. L. Kruk, and R. M. Wightman, Eur. J. Pharmcol. *109:*341 (1985).

136. J. O. Howell, W. G. Kuhr, R. Ensman, and R. M. Wightman, J. Electroanal. Chem. *209:*77 (1986).

137. M. A. Dayton, A. G. Ewing, and R. M. Wightman, Anal. Chem. *52:*2392 (1980).

138. M. Fleischmann, F. Lasserre, J. Robinson, and D. Swan, J. Electroanal. Chem. *177:*97 (1984).

139. M. Fleischmann, F. Lasserre, and J. Robinson, J. Electroanal. Chem. *177:*115 (1984).

140. J. Heinze and M. Storzbach, Ber. Bunsenges. Phys. Chem. *90:* 1043 (1986).

141. P. Bindra, A. P. Brown, M. Fleischmann, and D. Pletcher, J. Electroanal. Chem. *58:*31 (1975).

142. P. Bindra, A. P. Brown, M. Fleischmann, and D. Pletcher, J. Electroanal. Chem. *58:*39 (1975).

143. A. Russell, K. Repka, T. Dibble, J. Ghoroghchian, J. J. Smith, M. Fleischmann, C. H. Pitt, and S. Pons, Anal. Chem. *58:*2961 (1986).

144. J. W. Bixler and A. M. Bond, Anal. Chem. *58:*2859 (1986).

145. S. B. Kho, H. Gunasingham, K. P. Ang, and B. T. Tay, J. Electroanal. Chem. *216:*115 (1987).

146. F. Belal and J. L. Anderson, Analyst *110:*1493 (1985).

147. D. E. Weisshaar, D. E. Tallman, and J. L. Anderson, Anal. Chem. *53:*1809 (1981).

148. T. E. Mallouk, V. Cammarata, J. A. Crayston, and M. S. Wrighton, J. Phys. Chem. *90:*2150 (1986).

149. J. B. Justice (ed.), *Voltammetry in the Neurosciences,* Humana, Clifton, N.J., 1987.

150. J. A. Stamford, J. Neurosci. Methods *17:*1 (1986).

151. A. S. Baranski and H. Quon, Anal. Chem. *58:*407 (1986).

152. A. S. Baranski, Anal. Chem. *59:*662 (1987).

153. J. D. Genders and D. Pletcher, J. Electroanal. Chem. *199:*93 (1986).

154. R. C. Engstrom, M. Weber, D. J. Wunder, R. Burgess, and S. Winquist, Anal. Chem. *58:*844 (1986).

155. E. W. Kristensen, R. L. Wilson, and R. M. Wightman, Anal. Chem. *58:*986 (1986).

156. L. A. Knecht, E. J. Guthrie, and J. W. Jorgenson, Anal. Chem. *56:*479 (1984).

157. J. G. White, R. L. St. Claire, and J. W. Jorgenson, Anal. Chem. 58:293 (1986).

158. J. G. White and J. W. Jorgenson, Anal. Chem. 58:2992 (1986).

159. M. Goto and K. Shimada, Chromatographia 21:631 (1986).

160. H.-Y. Liu, F.-R. F. Fan, C. W. Lin, and A. J. Bard, J. Am. Chem. Soc. 108:3838 (1986).

161. R. B. Morris, D. J. Franta, and H. S. White, J. Phys. Chem. 91:3559 (1987).

162. C. E. Chidsey, B. J. Feldman, C. Lundgren, and R. W. Murray, Anal. Chem. 58:601 (1986).

163. T. V. Shea and A. J. Bard, Anal. Chem. 59:2101 (1987).

164. B. J. Feldman, A. G. Ewing, and R. W. Murray, J. Electroanal. Chem. 194:63 (1985).

165. B. J. Feldman and R. W. Murray, Anal. Chem. 58:2844 (1986).

166. A. G. Ewing, B. J. Feldman, and R. W. Murray, J. Electroanal. Chem. 172:145 (1984).

167. L. Geng, R. A. Reed, M. Longmire, and R. W. Murray, J. Phys. Chem. 91:2908 (1987).

168. C. E. D. Chidsey and R. W. Murray, Science 231:25 (1986).

169. M. S. Wrighton, Science 231:32 (1986).

170. G. P. Kittlesen, H. S. White, and M. S. Wrighton, J. Am. Chem. Soc. 107:7373 (1985).

171. E. P. Lofton, J. W. Thackeray, and M. S. Wrighton, J. Phys. Chem. 90:6080 (1986).

172. R. Deitz and M. E. Peover, J. Mater. Sci. 6:1441 (1971).

173. T. E. Edmonds, Anal. Chim. Acta 175:1 (1985).

174. E. W. Paul, A. J. Ricco, and M. S. Wrighton, J. Phys. Chem. 89:1441 (1985).

175. G. P. Kittlesen, H. S. White, and M. S. Wrighton, J. Am. Chem. Soc. 106:7289 (1984).

176. T. Hepel and J. Osteryoung, J. Electrochem. Soc. 133:752 (1986).

177. R. Kelly and R. M. Wightman, Anal. Chim. Acta 187:79 (1986).

178. C. Guminski, H. Roslonek, and Z. Galus, J. Electroanal. Chem. 158:357 (1983).

179. J. Golas, Z. Galus, and J. Osteryoung, Anal. Chem. 59:389 (1987).

180. M. Ciszkowska and Z. Stojek, J. Electroanal. Chem. 191:101 (1985).

181. P. M. Kovach, A. G. Ewing, R. L. Wilson, and R. M. Wightman, J. Neurosci. Methods 10:215 (1984).

182. F. Gonon, M. Buda, R. Cespuglio, M. Jouvet, and J.-F. Pujol, Nature 286:902 (1980).

183. E. Theodoridou, J. O. Besenhard, and H. P. Fritz, J. Electroanal. Chem. 122:67 (1981).

184. E. Theodoridou, J. O. Besenhard, and H. P. Fritz, J. Electroanal. Chem. 124:87 (1981).

185. T. E. Edmonds and J. Guoliang, Anal. Chim. Acta 151:99 (1983).

186. A. C. Michael and J. B. Justice, Anal. Chem. 59:405 (1987).

187. C. Kozlowski and P. M. A. Sherwood, J. Chem. Soc. Faraday Trans. 1 80:2099 (1984).

188. P. M. A. Sherwood, Chem. Soc. Rev. 14:1 (1985).

189. M. R. Cushman, B. G. Bennett, and C. W. Anderson, Anal. Chim. Acta 130:323 (1981).

190. J. P. Sottery and C. W. Anderson, Anal. Chem. 59:140 (1987).

191. J. Golas and J. Osteryoung, Anal. Chim. Acta 186:1 (1986).

192. J. Golas and J. Osteryoung, Anal. Chim. Acta 181:211 (1986).

193. G. Schulze and W. Frenzel, Anal. Chim. Acta 159:95 (1984).

194. V. J. Jennings and J. E. Morgan, Analyst 110:121 (1985).

195. A. Hussam and J. F. Coetzee, Anal. Chem. 57:581 (1985).

196. H. Huiliang, C. Hua, D. Jagnor, and L. Renman, Anal. Chim. Acta 193:61 (1987).

197. D. R. MacFarlane and D. K. Y. Wong, J. Electroanal. Chem. 185:197 (1985).

198. Y.-T. Kim, D. M. Scarnulis, and A. G. Ewing, Anal. Chem. 58:1782 (1986).

199. A. Meulemans, B. Poulain, G. Baux, L. Tauc, and D. Henzel, Anal. Chem. 58:2091 (1986).

200. M. Armstrong-James, K. Fox, and J. Millar, J. Neurosci. Methods 2:431 (1980).

201. N. Sleszynski, J. Osteryoung, and M. Carter, Anal. Chem. 56:130 (1984).

202. M. Fleischmann, J. Ghoroghchian, and S. Pons, J. Phys. Chem. 89:5530 (1985).

203. M. Fleischmann, J. Ghoroghchian, D. Rolison, and S. Pons, J. Phys. Chem. 90:6392 (1986).

204. H.-J. Huang, P. He, and L. R. Faulkner, Anal. Chem. *58*:2889 (1986).

205. A. M. Bond and P. A. Lay, J. Electroanal. Chem. *199*:285 (1986).

206. J. W. Bixler, A. M. Bond, P. A. Lay, W. Thormann, P. Vanden-Bosch, M. Fleischmann, and B. S. Pons, Anal. Chim. Acta *187*:67 (1986).

207. W. Thormann and A. M. Bond, J. Electroanal. Chem. *218*:187 (1987).

Author Index

Numbers in parentheses are reference numbers and indicate that an author's work is referred to although his name is not cited in the text. Underlined numbers give the page on which the complete reference is listed.

Abraham, M. H., 10(84,93), 11 (84), 12(84,93), 20(84,86,87, 88,89,90), 54(90), 65(93), 105 (93), 117, 118(86,87,88,89,90), 135
Abyaneh, M. Y., 239,241(117), 263
Adams, R. N., 266(2), 342
Akimoto, K., 287(46), 289(47), 344
Albani, O. A., 198(66), 261
Albery, J., 280(20), 343
Alemayehu, B., 92(191), 139
Alemu, H., 89, 90, 91, 92(184), 139, 93(197), 140, 94(199), 140, 118(A23), 129
Alexander, J. J., 108(217), 141
Alexander, R., 17, 18(80), 134
Allara, D. L., 220(146), 264
Alleman, E., 5(20), 132
Amatore, C. A., 267(6), 342, 294, 295(66), 344, 296(70,71), 345, 297(71), 345, 299(78,99), 345, 310, 313(111), 346

Amblard, G., 72, 73, 79(158), 138
Ambrose, J. R., 248, 249(125,126), 263
Anderson, C. W., 335(190), 350
Anderson, J. L., 301(86,87), 345, 301(89,90), 346, 321(146,147), 348, 326(90), 346, 335, 339 (146), 348
Andre, J. M., 204(132), 264
Andreev, V. N., 25(113), 136
Ang, K. P., 321(145), 348
Antonie, J. P., 10, 12(92), 135
Aogaki, R., 116(A16), 129
Aoki, K., 283(30), 284(32,37), 343, 287(46), 289(47), 291(58, 59,60,61,62), 344, 293(58), 344, 294(58), 344, 294(69), 345, 296 (69), 345, 298(37), 343, 305(96, 97,98), 346, 335(96), 346
Armstrong-James, M., 339(200), 350
Arndt, D. P., 175(40), 260
Arvia, A. J., 198(66), 261
Arwin, H., 203(75), 205(75), 261

355

Subject Index

Printed and bound by CPI Group (UK) Ltd, Croydon, CR0 4YY

23/10/2024

01778237-0015